MW00675804

THE MILL

RADE B. VUKMIR, MD, JD

University Press of America, Inc.
Lanham • New York • Oxford

Copyright © 1999 by
University Press of America,® Inc.
4720 Boston Way
Lanham, Maryland 20706

12 Hid's Copse Rd.
Cumnor Hill, Oxford OX2 9JJ

Library of Congress Cataloging-in-Publication Data

The mill / (edited by) Rade Vukmir.
p.   cm.
Includes bibliographical references and index.
1. Jones & Laughlin Steel Corporation—History. 2. Steel industry
and trade—United States—History. 3. Iron industry and trade—
United States—History. I. Vukmir, Rade.
HD9519.J64M55   1999   338.7'669142'0973—dc21   99—26069 CIP

ISBN 0-7618-1415-9 (cloth: alk. ppr.)

Contents

Acknowledgments   vii
Prologue   ix
From the Author   x

Section I: Prologue

Chapter 1:   Introduction   2
Chapter 2:   The Mill: *Jones and Laughlin Steel Company,*
             Aliquippa, PA   1909-1990   17

Section II: North Mill

Chapter 3:   Byproducts   34
             Fred Mancini   34
             Ed Reed   42
             James Byrd   47
Chapter 4:   Steelworks   54
             Michael Teleha   54
             Patrick A. Darroch   65
             David Henderson   74
Chapter 5:   Blast Furnace   90
             Kazimierz Pudyh   90
             Francis Yakich   100
Chapter 6:   Blooming Mill   113
             Eli Matish   113
             Joe Biss   119
Chapter 7:   Strip Mill   127
             Marion Prajsner   127

Section III: South Mill

Chapter 8:   14 Inch Mill   140
             James R. Coe   140
             Dennis Frioni   147
Chapter 9:   Welded Tube   155
             Elmer Cumberledge   155
             John Turkovich   159
             Anthony Rivetti   162
             James Downing, Jr.   166
             Bob Jurasko   175
Chapter 10: Seamless Tube   189
             Louis Salvoldi   189
             Alex Dudak   194
             John Kennedy   198
Chapter 11: Tin Mill   209
             George Ferezan   209
             John Carr   217
Chapter 12: Rod and Wire   226
             Michael Rebich   226

Section IV: General Maintenance

Chapter 13: Boiler Shop/Pipefitters   243
             Martin Zelenak   243
Chapter 14: Brick Layers   255
             Albert Awad   255
Chapter 15: Carpenter Shop   261
             Carl Ross   261
Chapter 16: Electricians   271
             Don Inman   271
             Lelio Ianessa   280
Chapter 17: Garage   285
             Larry Mitko   285
Chapter 18: Labor   294
             Guido Colona   294

Chapter 19: Machine Shop   300
            Joe Perriello   300
            Chris Trotta   314
            Leni Vukmir   320
Chapter 20: Mechanical   323
            Ronald D. Zuccaro   323
Chapter 21: Riggers   330
            Mitch Vignovich   330
Chapter 22: Electric Crane   339
            Dan Kosanovich   339

Section V: Service

Chapter 23: Clerical   348
            Ann Baljak   348
Chapter 24: Union   353
            Melvin Kosanovich   353
            Richard Vallecorsa   363
Chapter 25: Management   374
            John Cindrich   374
            Bill Stephans   378
Chapter 26: Security   386
            John Sudar   386
Chapter 27: Health Center   395
            Janice A. Paul   395
Chapter 28: Railroad   402
            George Stokes   402
            Steve Palambo   409

Section VI: The Final Chapter

Chapter 29: Historyof Aliquippa   415
Chapter 30: Epilogue   421

References   428
Index   429

The author's grandfather wrote the following letter to his newspaper's editorial page. He came to this country at 15 years of age and learned English by studying the dictionary.

Worker Has Suggestion

I, like the other 80 percent workers on the strike, cannot understand why we would not accept the company's offer in hope to obtain a new agreement. My impression is, our company is willing to give us better pension and insurance this year, and next year increase the wages. As we all know next year is not too far away. I think it is nothing to be ashamed of to go back to work. Because our economical problem is more important to us than saving the face of our leaders. I respect our union leaders very much, but as we all know, we cannot live on that respect, so we should understand what is possible and what is not possible. Therefore, my fellow men on the strike, I urge you to have a union meeting of all members immediately and in that meeting we should vote whether or not we should accept the company's offer. Therefore, let the majority decide on that so important issue to us. I say again we should accept the company's offer. I say this not b e c a u s e I am in economical trouble myself, but because it is a benefit to all of us; we all know this. Our union officials should go down to Wheeling with best hope to find something for agreement.

that business people of this country have plenty of difficulty in the competitive field, because of low cost foreign products. Everyone can see that everyday. It is very important to our country to preserve our long leading standard of living in the world, and to preserve that lead it is necessary to preserve the buying power at home. Therefore, both sides should work on a give and take basis.

Therefore, I, man on strike, wish to make the following motion regarding our strike: We steel workers should ask for 32 hours work week. but get paid for 36 hours. Probably the company does not wish to carry 4 hours pay, therefore, the union should give the same concession such as shift benefits: 3-11; 11 to 7; Sunday work percent, and overtime work benefits. Work 8 hours, pay for 8 hours, not for 12 hours: only pay time and a half over 40 hours a week. Cost of living should remain as it was in the last contract, so the union should pay for 2 extra hours from above said benefits, as before and company for 2 hours. Incentive both sides should have equal right on the decision.

Senority always has been and therefore should be now. better

We union steel workers know very well that the problem before us is nowhere; which is to say far from settlement. Everyone knows, who wish to know, that steel workers cannot be without work forever; neither can the steel mills be shut down forever. Therefore an agreement must be found sooner or later. Why not find it now? In this strike everybody loses, from country to steel mill sweepers. All that is needed is a little good will and mutual understanding on both sides. Many of us Union members know, very well, that 80 percent of the steel workers were satisfied with the last contract. In the mill we have been talking about why raise the wages now, because it is well known that the same workers in this Wheeling Steel Mill was making from $5 to $11 per hour; and if we get the same raise that raise will not cover our loss in this strike. It is very easy to solve the problem if any sincerity exists among union and company chiefs.

My dear fellow workers whether we like it or not, we must admit Senority always has been and therefore should be now, better and responsible worker should have the job in accordance with his ability. Because in last contract senority condition was plenty of a mess; everyone knows that. As far as Blast Furnace is concern, I know that too well.

I sincerely hope this plan will be acceptable to both sides for negotiation on a give and take basis. If it is not acceptable then our chiefs from both sides should find something better. I am certain the great majority of workers on strike will decide in their own favor. I think for us it is much better to accept that offer and go work 8 hours a day; than to humiliate ourselves by standing in line for 8 hours. Therefore, we could have the best 10 year contract with Wheeling Steel Corp., if approved from both sides by the officials; and why not approve it?

And now my sincere regards to all of you fellow workers.

RADE BEUKOVIC
2650 Cleveland Ave.

(Editors Note: Letters to the editor are welcome with few restrictions except that no letter will be printed if it is not signed. Letters should be confined to the point and should not exceed 150 words—less would be better.)

Acknowledgments

This work would not have been possible without the time and consideration of:

Mr. Donald Inman, President — Center for the Industrial Heritage of Beaver County
Mr. Inman almost single-handedly, carefully gathered and protected a significant portion of the memorabilia that exists concerning this steel mill in the early 1980s. His foresight and vision should be recognized and appreciated by those who view this information.

Mr. Joseph Perriello, President — Steelworkers Organization of Active Retirees
Mr. Perriello, an active union organizer in the mill's heyday continued on into retirement, galvanizing a group of steelworker retirees into an effective, caring, and political organization that has direct impact on the lives of former steelworkers, their families and Aliquippa residents. Mr. Perriello was instrumental as an intermediary between the Steelworkers Organization of Active Retirees and myself.

Melvin Kosanovich, General Griever — United Steelworkers 1211
Mr. Kosanovich, is an energetic union organizer whose area of expertise is arbitration and dispute resolution. His enthusiasm for the cause enabled us to recruit current steelworkers to discuss the issues of their day.

Gary Guydosh, Director — Gary Guydosh Photographic Services
Gary's enthusiasm for the project and keen photographic eye provided the vast majority of cameo and geographic photographic scenes in this work. The hours spent and his interest are obvious in the quality of the photography.

Dr. David Wollman, Chairman — Geneva College, Department of History

Dr. Wollman's foresight along with Mr. Inman, created a museum site within the Geneva campus, The Center for the Industrial Heritage of Beaver County, so that the information and memorabilia concerning the Jones and Laughlin plant could be archived and available to the public.

Mary Columbo, Director — BF Jones Memorial Library

Ms. Columbo was my first contact for background research concerning the Jones and Laughlin steel plant. A great deal of early community photographs were provided by this institution in a benevolent fashion.

Christine Henderson, Administrative Secretary — University of Pittsburgh Medical Center

This work would have literally been impossible without Christine's time, effort, and good nature concerning the multiple changes and rewrites that took place. Her editorial eye was helpful to soften the rough and sometimes harsh edges of this work.

The Steelworkers — Aliquippa

There would be no book without your remembrances and stories.

Photographic Credits:

The featured photographs were obtained from public record sources composed by these photographers. These photographs were entirely reproduced or composed by Gary Guydosh of Gary Guydosh Photographic Services.

Sheffield Studios — Nick Ferri
Associated Photographers
Goll Photography
Howard Earl Day Photography
Newman Schmidt

Prologue

The stories represented and portrayed to you were offered by the steelworkers themselves. Their opinions and thoughts are felt by themselves to be true and are representations of opinion whose validity can not be verified.

The memories presented are both spontaneous and the result of direct questions.

From the Author

This is the story of steel, of family, and of community, Aliquippa, Pennsylvania. The steel industry allowed simple, hard working people to survive and live for the day that their children could do one better than they could.

For those of you who will never know, since the gargantuan mills are gone, the testimonies of these men are recorded. This is their story. You would be afraid to walk where they toiled day-to-day their whole lives at the risk of their health, either acutely or chronically, just to get by. Steelworkers of J&L Aliquippa, at one point in time, comprised the largest integrated steel making plant in the world.

The mill itself no longer exists but go stand on the site of the Ohio and you will feel the memories and the presence of those who have come before you . . . I have.

Do not go gentle into that good night,

Rage, rage against the dying of the light.

— Dylan Thomas

Section I

Prologue

Chapter 1

ೞೞ

Introduction

When I was five years old I would walk from my grandparents house, up the hill from 2666 Cleveland Avenue, to Sunset Boulevard. There I would wait for my "Jedo" or grandfather to return from a days work in the mill. He was a large, hulking man, six feet four inches, 240 pounds, but gentle in feature and in form with bushy gray hair and a powerful gait. I can not remember for my lifetime, him ever raising his voice or his hand in anger.

There I would wait for the bus to return my grandfather home from the daylight shift at Wheeling Steel Works in Steubenville, Ohio. He came to this country at sixteen years of age. He did not speak a word of English nor did he ever learn to drive in his seventy-two years. He would lumber off the bus, he would extend his hand, grasping mine, saying hold my hand, it's time to go home.

My grandfather worked for fifty-five years as a watertender on the blast furnace. It was hard, dirty work, where men such as he handled 250 pound brass fittings to the water circulation system that kept the furnace cool. They worked in small and tight areas where machinery could not be taken to make the job easier. These men toiled through the day-to-day operations of keeping the furnace cool, and their workload increased during furnace shutdown when these pipe systems were repaired.

I remember that my grandfather retired when he was sixty-eight years old, and two weeks after retirement the mill called him back, stating that they would like to know if he could return to work past the mandatory retirement age of sixty-eight. He gladly returned and was employed until he was seventy-two years of age. Men such as he had knowledge that could not be contained in blueprints or in plans or by young engineers fresh out of college.

I later found out that the reason for the call back was that during a furnace shut-down, without his knowledge of the older pipes of the system, a good proportion of these could not be located by the younger personnel. When he went back to the site to be questioned by young engineers, who said it wasn't on their plans or blueprints, he would confidently tell them to dig under the foot of ash and molten irons that lay on the floor. Sure enough, the coupling and fitting was found.

There is one other story about my grandfather that was related to me later in life. At one point, he came home from work and my grandmother, Anka Beukovic, had the table set for dinner. This wonderful lady, a substantial woman with long black hair kept in a neat bun, was very much remembered as being a good-hearted person and one of the best cooks in the small steel town she had offered many meals to those less fortunate.

She had prepared the evening dinner for my grandfather on his way home from work. He declined the dinner offer, said he had a headache and that he was going directly to bed. My grandmother reluctantly agreed and said that she would keep things warm if he wanted to eat later.

After one hour there was a knock on the door and one of the family friends had asked my grandmother how her husband, Rade Beukovic, was doing, and she said that he was fine and that he had just gone to sleep with a headache. This man proceeded to tell my grandmother that there had been a carbon monoxide leak at the mill where my grandfather was working. He had accepted this fact and carried nine men to safety and fresh air, away from the furnace gas. His return at home, however, had occurred without fanfare, and I remember him to be a quiet, understated man. He taught me the value of education and learned English by reading the Webster dictionary, ten words, each night, from cover to cover — a practice he continued for as long as I can remember.

My father, Matthew Vukmirovic, a neatly kept man with slicked back hair and well defined heavy features, was a rigger at the Jones and

Laughlin Aliquippa plant. This job title carried respect as these were men who erected and disassembled the various structural components and buildings of the mill. Their jobs were dirty, hot and dangerous. My father was also a quiet man that loved the outdoors and taught me more than he will ever know about how to be a good father. When I was thirteen years old, I got a call at school after some hushed discussions concerning the fact that a man had died at J&L during the night. That man was my father.

When I was eighteen years old I went to work in the mill during my summer vacation. My first summer was spent in the Blooming Mill. Work was hard but the company was good, and I remember a discussion that occurred around the lunch area one day. One of the workers was telling a story of a rigger who was dismantling a crane in the Bar and Billet area out back. He was using a burning torch and dark burning glasses, which prohibited him from seeing as he cut through the last supporting bolt that held the overhead crane. The safety man, as they were called — for whatever reason they will only know — did not see this either. As this man burned through the last bolt, the crane collapsed and my mother became a widow.

The next three summers I spent in the Blast Furnace Department learning a different aspect of the trade, and I worked a number of differ-

Rigger picnic (1970's)

Father Matthew Vukmirovic, Korean conflict

ent jobs: stock house, sinter plant, labor gang, blast furnace floor. During this time my mother, Leni Vukmir, a hard working woman whose worn hands and body could not belie her effervescent spirit and had to reenter the work force late in life to make ends meet. She had worked in the Tin Mill at the Weirton Steel Plant as a tin flopper during World War II and, as most women of that day, retired to raise two kids including my sister Alana. She went to work at the North Mill Machine Shop as a laborer, shoveling and cleaning up metal chips.

It made me uncomfortable, but later I realized the burden that she had had and the extent to which I am truly indebted to this wonderful woman. She similarly stressed the value of education and continually tried to improve us, as she would strive for educational opportunities for both my sister and I. I remember vividly a drive three days a week from Center Township to the Buhl Planetarium on the North Side, where I learned about the wonders of science for hours on end. My mother sat at Isleys or walked the Allegheny Center Mall awaiting our return. She was forced into retirement when the mill shut down.

Site of overhead crane collapse in Bar and Billet Mill

Overhead Crane

Blast Furnace water tender

Wedding day, grandparents — Rade and Anka Beukovic

In the summer of 1977 in Western Pennsylvania there were approximately thirty of us sitting in a room, in June, filling in the information for identification cards — including name, social security number, height, and weight, on a deep robin's egg blue colored card — a shade of color I have not encountered since then. We sat in rows and in columns, in grammar school desks, as if we were enrolling in class. People appeared to be nervous, some were joking, faking enthusiasm or bravado, some knew they had good, safe jobs waiting, privileges extended to friends and families of bosses or foremen. Others were just afraid. Sons and daughters of laborers didn't know what to expect. The kids talked about sports, cars, and girls mainly, but eagerly awaited what came before them.

It became quiet as an individual walked into the room and began to address the group. Slowly we all began to listen, a varied group, white and black, young men and women lucky enough to have landed a summer job at J&L. TD Duckworth, a small but imposing man, wore short sleeves, starched white shirt, simple patterned necktie, and a prominent sense of purpose. He welcomed us in a not overly friendly manner but described the pathways opened to us.

There were those who would truly go on to college as well as those who would never leave the town working as their fathers and mothers did. All I remember from that conversation was the adage that being employed at J&L was a two-way street: "an honest day's work for a good day's pay", and I remembered that.

Duckworth left, then a meekish man, Augie Palumbo, with a prototypical clerk appearance carrying a clipboard began to read our names. For each name that was read a department was assigned — Seamless Tube, Steel Works, Byproducts, Tin Mill. The Tin Mill processed cold steel in a clean environment. Byproducts was a facility where even the strongest men could not feel untainted by toxic byproducts of steel production. The Blast Furnace and Steel Works was as close an incarnation of hell on earth. My own lot cast me to a department called the Blooming Mill, and I waved the rest of my friends good-bye as we all headed off to separate departments.

They gave a tour of the facility quickly on a refurbished school bus and then dropped us off to the department. People said I was lucky, my boss, Nick Cavoulas, with wavy gray hair, short sleeves, plaid shirt and a casual manner, was nice and fair. The clerk that had been assigned to us traipsed us through the wash house as a group of 30 men congregated among a group of green benches clustered around a woodburning stove and discolored old refrigerators. The men looked up from lunch, dirty and tired. Some joked but most didn't say much. The annual migration of college kids was like the swallows of Capastrano returning every summer. The older men didn't seem to mind it much, they were comfortable in their jobs making good money.

The younger men, a year or two out of high school, had found that the glow had already warn off. They thought the job would get them cars, boats, and a nice house. It had lost its glamour, they were resentful, and occasionally would let it show.

The clerk of the individual department gave us our job assignments for the week and the posted schedule had our names penciled in, days of the weeks, followed by a number. — eight meant the daylight shift, four meant the afternoon shift, and twelve meant midnight or night shift.

People weren't sure what to wear and blue jeans and sweat shirts were the dress for the uninitiated. However, the veterans wore green. Special elements of the workforce that were exposed to the high heat or combustible environments — the mill-wrights, riggers, and other trades that burned and welded — were given flame resistant clothing. It was a pale green shade by the same designer as the cards, as I have never seen that color, that particular shade of green, again. We were assured that good footwear was required, and this was not a point of debate.

We were a sundry collection of folks who were taken to a semi-trailer with a boot painted on the side and told we could buy our boots on credit to be taken out of our first paycheck. Those who didn't have the money thought this was a grand idea, but as their fathers and grandfathers had found out, this work-to-live strategy of compensation made advancement near impossible. But for me, at that time, the pay meant I would have some spending money on payday.

Most of us started on the daylight shift the next day. We reported to the wash house, the area of congregation where men drank coffee, joked, and got their assignments for the day. Each individual department had areas of production, clean-up, maintenance, all overseen by the sub-foreman or pusher. The sub-foremen elected a small group of individuals to do the day's tasks. These sub-foremen were not really salaried employees but paid slightly more money, but really were more part of the labor force than management. The general foreman of the department ran the whole operation and wore casual dress clothes, khaki pants, and a white shirt. We were given lockers, put our lunches in the refrigerators, and headed out for our day. It was loud, hot, humid, and the acrid smell of sulfur permeated the entire site. I don't remember what I did that first day, but I do remember leaving the mill trying to make sure that all the procedures were accomplished, time card punched as I walked to the gate past the dusty signs about protecting plant security and safety. They were reminiscent of a bygone era, speaking of the duality of the Axis Powers and taking steel secrets in 1944, compared now to guys taking home pockets of nails to build their houses.

I did various jobs that summer from time wasting operations called general labor where I swept floors starting at one end, sweeping diligently to get to the other end, and finding it completely dust covered again. Somehow it was easy to lose enthusiasm, although we always tried hard for that good day's pay. I think my first summer I made eight dollars an hour, a huge sum of money in those days.

I do remember the various jobs that I worked that summer, however. Crane following seemed to be a common task as loads of hot or cold steel lifted by huge cranes with ten to some seventy-five tons lift capability. The crane operators were an elite bunch with high paying, good jobs, in an air conditioned cabin. I remember one fellow operated the 45 Inch New Blooming Mill with spit shined, pointed cowboy boots and a belt buckle to match. I had never seen that combination in the mill or anywhere else for that matter.

Crane follower was a dangerous job, and involved helping wrap the load with nylon lift straps around various items and giving the cranemen signals to move the load. The irony was that you had to memorize the thirty types of crane signals from the dusty chart on the wall, as if the crane men with forty years experience would listen to anyone who had been there for a week. But you felt like you were part of the team, and they usually managed to deliver the load safely. However, loads could shift, straps could brake, and you never stood closer to the load than you had to.

One variant of this job required switching between a lift-harness and a bucket called a clamshell to clean out a sump thirty feet deep where waste water metal shavings and filings from the Blooming Mill rolls. It was murky, dark, steamy water that you would look into and yet not imagine how black it really was. I worked this job mainly on the night shift, sitting in four hour shifts and biding the time. A lot of individuals fell asleep. I tried my best to stay up, with all manner of amusement, like chewing tobacco and putting decals on my helmet.

There were two Blooming Mills, the 44, and the 45 Inch Mill where they would take an ingot of steel that had been cast in the Steel Works reheat it and drawn it back and forth over a long series of rolls, hundreds of feet long, drawn faster and faster from one end to another. It was literally earth shaking in its force, sound, light, and color. The steel would move fifty feet in seconds and then stop and reverse direction in a shower of sparks and a cacophony of sound. Gradually the ingots became slabs and then were drawn into bars and cooled in the slab yard for presentation to a Finishing Mill.

I remember the men, they were mainly nice, someone's father or brother, and they were mostly happy making a good wage by combination of bonus and incentive, a concept I still do not understand to this day. But it allowed these men to make 100 dollars a day in 1977, the equivalent of 400 dollars a day in 1997 at a conservative five percent appreciation.

I remember one fellow named Gomer. A stocky, solid man, distant from the pictures he carried of a young solider in olive drab fatigues. The men said he had been to Vietnam, and I had never met anybody that had been there. Workers having been through military service was common, but those who had ended up in Southeast Asia were unlucky and not as common. It seemed to be the sentiment of the outside world but not in the mill. These men were respected. Gomer had a coffee can filled with pictures of dead Vietnamese soldiers, Vietcong, and NVA regulars. Kind of matter of factly, a lot of them were dead. I never asked how they got that way. I was too young to know, but too afraid to ask. However, Gomer probably hadn't been much older than I was when he was 7,000 miles away in some jungle. There didn't seem to be many protesters in the mill. They had American flag license plates and bumper stickers that read "Be American, buy American" and "I'm a US Steel Worker and I drive an American made car" stuck to the back of their pick-up trucks.

I remember the men talking in the Blooming Mill one day while they sat around the lunch area. They spoke about accidents men were known to have. There was a subtle frailty of the human body and spirit compared to the resilience and permanence of steel. It was a force that you could sometimes control, but occasionally not, steel ravaged the human body and spirit, as it was prone to do in its uncontrolled condition.

Behind the Blooming Mill wash house was the Blooming Mill slab yard where the semi-finished slabs were made from ingots and cleaned of imperfections by men known as scarfers. They used torches to burn the imperfections from the surface of the steel and this was an attractive job. When the steel was done it was stamped and certified, and I had worked two weeks in this area. The incentive was good and men made a lot of money.

Men were telling a story one day of time past when on a rainy, cold three to eleven shift in January a rigger work detail was dismantling the overhead crane used in that slab yard. The story was told that individuals should be conscientious and, as the story goes, a labor crew of some type had cleaned the rails so that the rigger crew could see the areas in which they were. But apparently the rails were not cleaned well enough. A man of forty-six was burning through the rail. He worked hard, and had dark shielded welding glasses. He should had been accompanied by spotters who could have watched his progress, as this was invisible to him with the dark welding glasses. It was unclear what had happened, but no one realized the extent of his cutting and the crane collapsed. He fell thirty feet to the ground, striking a railroad tie, and was taken to Aliquippa Hospital where he died. That was my father. The duality of the mill was obvious. It gave life and took it away either acutely or chronically, because no one left unscathed. I worked in the Blooming Mill that one summer and went off to college, and then came back for more.

That next summer seems a blur but we were experienced mill veterans. We brought back our hard-hats from the year before with scars of battle, dirty, scratched, decals applied, like a real steelworker. We had our washed green pants, and millwright's jackets, and our boots that had been worn through one summer. I was assigned that summer to the Blast Furnace, as were some of my friends.

The Blast Furnace had a number of sub-departments that included the Stock House, which was an underground tunnel that ran in a line for approximately sixty feet on underground railroad tracks, and a larry car that would accept a furnace charge of coke, iron pellets, or limestone from an above ground bin that would drop the products through into the larry car. As it would stop at various bins along the way, taking various elements of charging compounds when the right mix was obtained, the car came to a halt and was dumped into a car that would run on a vertical dolly and track to the top of the furnace one hundred feet away.

My main job there was to clean up the spilled iron ore, coke, and limestone that fell from the larry cars, especially in the bottom of the pit where the skip or bucket would take the load to the top of the furnace. The worst fear was that the cable would break and the skip would cut loose, and that you would be left without a chance at survival. Again, I worked a lot of nights with a senior larry operator who drove the car, and the controller who ran the skip and decided on the proportion of load composition. We formed a team of three, and I remember that we would buy coffee literally every hour for eight hours from the vending machine, and play poker hand with the cups.

I also spent time that summer in the Sinter Plant, which was an area that took iron making byproducts, particulate dust of coal, and iron ore that would be repackaged as sinter (which was compressed) a combination byproduct that could be recharged into the furnace to improve efficiency, but claiming as much excess raw material as possible. The entire sinter plant assumed a red rich glow like Georgia topsoil. This was indeed the dirtiest place in the world. Everytime you would strike a beam or a wall you would be left with this red, indelible discoloration, like war paint. This was a truly lethal inhalant mixture of articulate matter, and the weight of fiberoptic lung disease was high in these men.

If you wore the respirator you were protected somewhat, but you were working so hard you couldn't breathe, so most respirators got cast by the wayside. In the heat of summer, for days and even weeks after you left your sinter plant duty, you would continue to leech red material from your pores so that you could not wear a white dress shirt because of discoloration. These workers were mainly foreign, too afraid to complain. They would mutter through their day with thick accents, hoping to send their kids to college and pay off their house. Having worked there, I'm sure their health suffered.

I did an additional stint that summer on loan to another department called the 14 Inch Mill, the home of the junior I-Beam, a real money maker for the Jones and Laughlin Plant. The beam was a structural component for lightweight modular housing like trailers and it was very successful. The 14 Inch Mill was between the North Mill, making iron and steel, and the South Mills where completed steel products were made. The 14 Inch Mill sat along the river and most of its activities were manufacturing, moving, and storing slab steel products made from billets sent from the Blooming and Bar and Strip Mills. These finished products included angle iron used for guard rails, tire rims, and structural beams. Although cleaner, this was also a hot, dangerous place where men labored under the constant fear of shifting loads and being crushed by tons of steel as they hooked up cranes.

One other thing I remember was the volume of conversations in local Aliquippa neighborhoods. Most men were deaf, having sensory-neural hearing loss from constant noise exposure from places like the 14 Inch Mill. It was so loud — hundred decibels on a day-to-day basis — that most men were literally deaf by the time they reached middle age. So the usual conversation at family gatherings or in the middle of the mall was usually shouted at levels much to high for plain conversation. In fact, as the old steel workers had died off they were replaced by new, young commuters. I see the interface in the shopping mall. The old fellows yell at each other to be heard as young moms hurry their children along so they are not exposed to these loud old men.

My three months ended uneventfully as another summer had drawn to a close. As a college student I was young, strong, and on top of the world. Working the afternoon was the preferred shift, as opposed to the men who wanted to be with their families. College students would work until eleven, shower, and then head out to the night clubs until two a.m. to wake up late and return to work again.

My last two summers I worked in the Blast Furnace and transferred to the Labor Gang, an aggressive, highly varied job that had men troubleshooting various spots throughout the Blast Furnace. The Cast House floor was one of these places which was as close to hell on earth as you can find. These men, wearing layer upon layer of heat protective clothing even in the hottest summer, tried to protect themselves from the heating and burning of molten metal labored. They looked like spacemen wearing metallic flame resistant suits, like Neil Armstrong landing on a volcanic molten moon. Having previously sealed the outlet to the Blast Furnace, they "tapped the hole," allowing molten iron to run out hot and caressed with steel implements as if they were tending a flock of sheep directing them along the right course. The labor crew would be involved if there were spills or accidents which happened frequently.

I remember working for Bill Frost, the labor boss, with his bald head on which he wore a handkerchief, his full beard and gruff English attitude. Most of the workers thought he was crazy, but he seemed to respect hard work. I remember he summoned us once for a furnace explosion. There were fifty tons of molten iron left on the Cast House floor. He called us in and authorized overtime. I remember those days well and felt like we were really part of the team. They brought in huge cranes to help lift and clean, and small labor crews went in fifty and ninety pound jackhammers and bars to free the iron from the metal fittings. We worked hard for days. I remember my last day Bill sent me down with a crew of workers, mainly other college kids, but they were certainly not keen on the idea of cleaning up. It was hot, steaming, and very dangerous. A group of half a dozen went in for a bit, but all left. One or two said they were going to go home.

I said fine, I'll do it myself, went in, and worked for fifteen minutes. It was hot, steaming, and I began to feel a little bit woozy. I thought I had worked to hard or hadn't eaten enough and came out once, felt a little bit better, went back in a second time. At that point I got very dizzy, came out, and collapsed. As it turns out, this was a carbon monoxide exposure. I guess it was kind of a badge of courage for the place.

It was a good place to work in those days. Hard work was appreciated. Black and yellow industrial flashlights, were given to the hard workers, made them feel like part of team. I remember that same summer, my grandmother had died. I had been working hard, she was in the hospital, I remember talking to her and she couldn't hear me. She died later in the intensive care unit and was lasix at the time, too frail to survive more episodes of congestive failure.

During my time spent in the Blast Furnace, I had often gone to other areas. The cafeteria was also a right of passage, a reward for a double worked shifts. Sixteen hours in this environment, and you would go wait in line with the other men, get an unhealthy meal of two fried egg and cheese sandwiches premade, a bakery product, and a couple cans of juice. Younger fellows would stick these juice tabs to their helmets to show everybody how many doubles they had worked. During one of my excursions to this area, I encountered my mother who had worked in the Machine Shop as a laborer and clean-up person. She was a strong woman who did what she had to do to keep her family together. She tells me I had tears in my eyes when I saw her. I don't remember, I think I blocked that part of it out.

The reason I wrote this book was because the first time that I had driven along Constitution Boulevard, as I had done so many times before seeing the reassuring light of the mill in a sea of darkness, the blue eternal flame from a natural gas jet, and then the Christmas star on the Blast Furnace hundreds of feet above the ground every December. I drove home from college one day and there was a sea of black — no lights, no smoke, no nothing. It was silent and this way of life was gone forever. The work was profitable, the work was hard, but the company made money. However, the greed of the 1980's driven by Wall Street and the desire for shareholder and corporate profits had shut the mill down and ended a way of life forever. Hopefully, you will understand their stories, the steelworkers of Aliquippa, Western Pennsylvania, and the industrial Northeast.

United Steelworkers of America, Steelworkers Organization of Active
Retirees (SOAR)

Author — Rade B. Vukmir

Chapter 2

ഇറ

Jones and Laughlin Steel Company, Aliquippa, PA 1909-1990

An Honest Day's Work For

A Good Day's Pay.

T.D. Duckworth
1977

A Chronological History of
Jones & Laughlin Steel Corporation

1853   Partnership formed between John and Bernard Lauth, B.F. Jones, and S.M. Kier for a new iron works, called Jones, Lauth & Company, it was to be headquartered in Pittsburgh.

1856   John Lauth sold his interest in J&L to the other partners. James A. Laughlin bought part interest in the firm.

1859   James Laughlin, his two sons, B.F. Jones, his two brothers, and Richard Hayes formed Laughlin & Company. They erected blast furnaces and called them the Eliza furnaces — the name present furnaces bear.

1859   Cold rolling of steel bars patented by Bernard Lauth.

1860   The United Sons of Vulcan, a labor organization, was established in Pittsburgh and recognized by J&L.

1861   Bernard Lauth retired, James Laughlin bought his interest, and the company consolidated under the name Jones & Laughlin's, American Iron Works.

1862   Company built new mills and furnaces to meet Civil War needs.

1865   J&L's cold rolled steel becomes world famous. In addition, the United Sons of Vulcan adopted Jones' "sliding" scale of wages.

1881   The American Federation of Labor was founded at a Pittsburgh convention.

1883   Company became known as Jones & Laughlin's, Limited, and first stock was issued. Company also built first Bessemer converters, steam-powered blooming mill, added new bar mills.

1884   B.F. Jones, one of the founders of J&L, became president of the American Iron and Steel Association.

1894   J&L became large producer of cold rolled shafting, nails, bars, plates, rails, and sheets.

1895   J&L's first two open hearth furnaces are built at Pittsburgh. That same year, ore properties in the Great Lakes region were bought and leased by J&L, a key step in integration of the company.

1899   A new hot bar mill of revolutionary "looping" design is built at Pittsburgh, starting a new era in hot rolled bar production for industry.

1900   J&L's annual production goes up to 650,000 tons of steel ingots, more than was made in the entire U.S. when J&L started in 1853.

1902   The company is reorganized into Jones & Laughlin Steel Company with capital of $30 million.

1903   B.F. Jones, Sr., dies, Dean of Ironmasters. This year marks the end of the first great period of consolidation among the various iron and steel companies in the U.S..

1905   J&L purchases property 20 miles from Pittsburgh in Woodlawn, PA, on which it plans to build a steel mill, later to become the Aliquippa Works.

1906   J&L expansion continued, with erection of eight-story office building, blooming mill and launching of first two ore freighters of J&L's lake fleet, steamers "B.F.Jones" and "James Laughlin."

1907   First "safety first" program is started to reduce hazards in J&L mills.

1908   The tunnel leading to the Aliquippa Works is constructed.

1910   Tin mills start operation at Aliquippa as J&L becomes first integrated independent to challenge "tin plate trust." J&L enters the wire business, with completion of new rod and wire mills at Aliquippa Works.

1912   Steel is first produced at J&L's Aliquippa Works.

1916   J&L enters the tubular products business with the installation of welded pipe works at Aliquippa. An additional bar mill is constructed to help produce steel for war requirements.

1923   J&L incorporates under its present day name, Jones & Laughlin Steel Corporation, and sells its first stock on the open market.

1924   At Aliquippa, a 14 inch bar mill of unique construction to roll special shapes is put in operation.

1926   First of three seamless tube mills built at Aliquippa.

1928   Cold rolled steel sheet begins to replace hot rolled sheet in the making of tin plate.

1930   1500- Ton capacity blast furnace built at Aliquippa Works, first of its size in the industry.

1935   Thirty-four percent of J&L's total tonnage is in hot rolled and cold finished bars, one percent is in sheet or strip. J&L begins major product diversification in the 1930's, expanding into wire, tin, plate, welded and seamless pipe, cords, and wire rope.

1937   J&L enters sheet-strip business and erected its first hot strip mill for this operation.

1953   Major expansion and erection of new facilities to increase steel production is undertaken. At Aliquippa Works, a tin mill warehouse is completed and electric drive is installed on Blooming Mill.

1955   J&L acquires Monarch Steel Company, Hammond, Indiana.

1957   BOF shop is built at Aliquippa Works.

1966   J&L started construction of Hennepin Works in Illinois to meet increased steel demands for manufacturing purposes in the Midwest.

1968   Ling-Temco Vought, Inc. (LTV Corporation) acquired controlling interest in J&L.

1969   J&L becomes first to market high strength, low alloy steels.

1974   J&L becomes a wholly-owned subsidiary of the LTV Corporation.

1974   A $200 million expansion program for the Aliquippa Works is announced. The program, includes facilities to increase the output of the basic oxygen furnace shop, installation of a second blooming mill to provide sufficient semi-finishing capabilities, additional tubular products, finishing facilities, and improvements and additions to coke producing and iron-making facilities to support the added steel output.

1976   J&L becomes one of the seven largest steel producers in the nation, with operations in twenty-nine states. The diversified product line of the company includes: sheet mill products, coated sheet products, tubular products, conduit products, hot rolled bar and shape products, cold finished bar products, stainless, alloy and trip steels, tin mill products and industrial steel products.

1978   LTV acquires Youngstown Sheet & Tube.

1981   LTV acquires Republic Steel.

1982   Aliquippa shutdown begins, Rod & Wire Mill is shutdown.

1985   BOF, Blast Furnace, Byproducts shutdown.

1989   Aliquippa Works shutdown complete.

1989   LTV Tin Mill resumes operation.

1990   J&L Structural Steel resumes 14 Inch production.

Reference 1

Aliquippa Works Operations
1975 - 1979

Over 10,000 Employed

3 Coke Batteries

3 to 4 Blast Furnace Operation

3 Basic Oxygen Furnaces

Continuous Billet Caster

44 Inch and 45 Inch Blooming Mills

21 Inch and 18 Inch Billet Mills

44 Inch Hot Strip Mill

14 Inch Structural Mill

Welded Tube Department (Continuous & Electric Weld)

Seamless Tube Department

Rod & Wire Department

Tin Plate Department

30 Inch Round Mill

Aliquippa Works

Year	Decline	Steel Production	Active Employees
1957			14,800
1972			10,400
1975			10,600
Sept. 1979	A-4 Coke Battery	3,061,000 Tons	10,300
Sept. 1981	Rod & Wire Dept.	2,531,000 Tons	9,600
Oct. 1981	A-4 Blast Furnace		
Nov. 1981	45 Inch Blooming Mill		
Feb. 1982	A-1 Blast Furnace	893,000 Tons	9,700
March 1982	44 Inch Strip Mill		
April 1982	Seamless Tube Dept.		
1983		810,000 Tons	4,200
March 1984	30 Inch Round Mill	852,000 Tons	3,900
Sept. 1984	A-5 Coke Battery		
Oct. 1984	44 Inch Blooming Mill		
Oct. 1984	21 Inch & 18 Inch Billet Mills		
Jan. 1985	Welded Tube Dept.	174,000 Tons	2,800
June 1985	Coke Plant		
July 1985	Blast Furnace Dept.		
	B.O.F. & Continuous Caster		
July 1985 operating	14 Inch Structural Mill		
operating	Tin Plate Dept.		
1986			800
1990	Blast Furnace demolition begins		
April 17	A-4 (Built 1912)	4:14pm	
April 24	A-5 (Built 1919)	4:28pm	
May 11	A-1 (Built 1909)	2:55pm	
May 15	A-3 (Built 1910)	3:35pm	
May 23	A-2 (Built 1910)	6:21pm	

Wage Scale (1947)

Job Class	Standard Hourly Wage Rate	Job Class	Standard Hourly Wage Rate
0-1	$ .965	16	$1.49
2	1.00	17	1.525
3	1.035	18	1.56
4	1.07	19	1.595
5	1.105	20	1.63
6	1.14	21	1.665
7	1.175	22	1.70
8	1.21	23	1.735
9	1.245	24	1.77
10	1.28	25	1.805
11	1.315	26	1.84
12	1.35	27	1.875
13	1.385	28	1.91
14	1.42	29	1.945
15	1.455	30	1.98

United Steelworkers Of America
Membership Oath

"I do sincerely promise, of my own free will, to abide by the laws of this Union; to bear true allegiance to, and keep inviolate the principles of the United Steelworkers of America; never to discriminate against a fellow worker on account of creed, color, sex, or nationality; to defend freedom of though, whether expressed by tongue or pen, to defend on all occasions and to the extent of my ability the members of our organization."

"That I will not reveal to any employer or its agent the name of anyone who is a member of our Union. That I will assist all members of our organization to obtain the highest wages possible for their work; that I will not accept a member's job who is idle for advancing the interests of the Union or seeking better remuneration for the member's labor; and, as the workers of the entire country are competitors in the labor world, I promise to cease work at any time I am called upon by the organization to do so. And I further promise to help and assist all members in adversity, and to have all workers join our Union that we may all be able to enjoy the fruits of our labor; that I will never knowingly wrong a member of see the member wronged, if I can prevent it".

"To all this I pledge my honor to observe and keep as long as life remains, or until I am absolved by the United Steelworkers of America."

Lets stand for the flag salute:

I pledge allegiance to the flag
of the United States of America
and to the republic for which it stands
one nation under God indivisible
for liberty and justice for all

and a moment of silence please

Steelworkers Organization of Active Retirees
June 25, 1996

United Steel Workers Union Local 1211
Aliquippa, Pennsylvania
Past Presidents

Paul Normile	–	02/13/37 – 06/30/40
Joseph Krivan	–	07/01/40 – 06/30/42
John Atkinson	–	07/01/42 – 06/30/46
Tom Breslin	–	07/01/46 – 06/30/48
Mike Pollick	–	07/01/48 – 06/30/50
Anthony Vladovich	–	07/01/50 – 06/30/58
Nicholas Mamula	–	07/01/58 – 07/20/61
Louis Biega	–	07/20/61 – 06/30/62
Morris E. Brummitt	–	07/01/62 – 06/30/67
Nicholas Mamula	–	07/01/67 – 06/30/70
Morris E. Brummitt	–	07/01/70 – 06/30/73
Louis DeSena	–	07/01/70 – 06/20/76
Alex George	–	06/21/76 – 05/16/79
Peter J. Eritano	–	05/17/79 – 06/20/85
Richard Vallecorsa	–	06/21/85 – present

Flood in Woodlawn Valley (June 26, 1909)

Franklin Avenue and State Street

Aerial view of the Aliquippa Works located at the Ohio River

West Aliquippa

The Mill

War effort

Aliquippa from Mead Street

Plant entrance

Jones & Laughlin Steel Corporation — Brief History

In the year 1853, Benjamin Franklin Jones bought an interest in the firm of B. Lauth & Brother, a small iron works located along the bank of the Monongahela River at Brownstown, Pennsylvania, now a part of the South Side Lauth & Company, American Iron Works. In 1854, James Laughlin obtained an interest in the company and upon retirement of Mr. Lauth in 1861, the firm's name was changed to Jones & Laughlin, American Iron Works. This was the beginnin of the company that was incorporated as the Jones & Laughlin Steel Corporation in 1923.

The Jones & Laughlin Steel corporation and Subsidiary Companies were engaged in all major functions related to the manufacture and distribution of steel, steel products and coal chemicals. As of January 1, 1960 the Corporation has approximately 44,00 employees and has an annual ingot capacity of 8,125,000 net tons, making it the fourth largest steel producer in the United States.

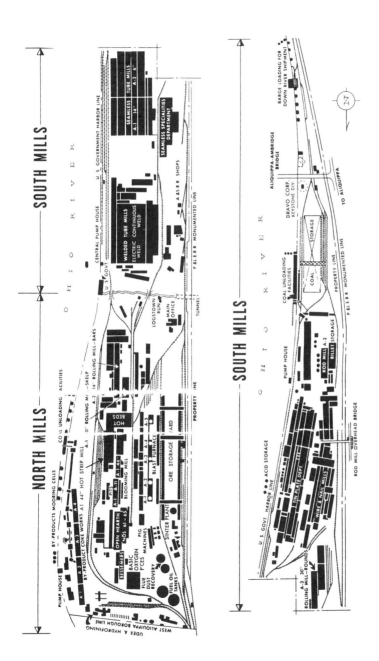

ALIQUIPPA WORKS DIVISION

JONES & LAUGHLIN STEEL CORPORATION

Reference 2

The citizens of Aliquippa express their appreciation

The mill across Constitution Boulevard

Section II

North Mill

Chapter 3

ℰᴏᴄ℞

Byproducts

H e was a quiet self-assured man, gray hair and quiet sleepy eyes, but as a young man he stood strong with thick forearms protruding from his short sleeved-checkered shirt.

My name is Fred Mancini from Monaca. I grew up in West Aliquippa, near the mill there. I was close to it. I'll be seventy on June 24th. Born and raised in West Aliquippa. Growing up in West Aliquippa was dirty, filthy. My mother got up in the morning, she had to sweep the porch and wipe the windows off, everyday.

Well, I worked from 1942 part-time and then joined the Navy. I came back out of the Navy and went back into the mill in 1947. Then I got called back in the Korean War, and when that was over in 1952, went back into the mill to the Byproducts, from 1952 until I retired. I didn't want to retire in 1983, they forced us. I had forty-one years of service but service, time didn't count. No, that was broken because I didn't go back before ninety days were up.

I'm not sure, in 1942 I think it was either forty-eight or seventy-eight cents an hour. I'm not sure, can't remember. But I know that I would work two days a week, like after football and on Saturday mornings. If you worked in the mill you didn't have to go to football practice.

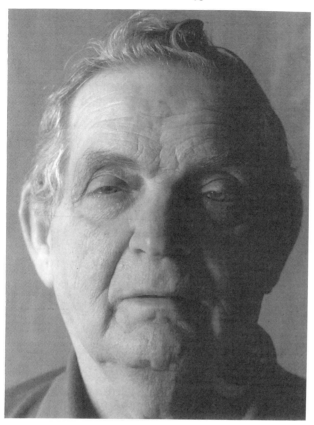

Fred Mancini

So I worked four days, and I know I brought home about eighteen dollars for two weeks for four days to my mother. I would give my mother the paycheck and she would give me three or four dollars and that was enough to keep me going for two weeks.

Well, there were ten us and everything helped, every little bit helped. My dad was sick from day one. He had diabetes. In them days nobody knew what to do, and there was ten of us. When the war started, my oldest brother joined the Navy. He had a job in the mill. Then I went in the mill, then when I graduated, I joined the Navy and then things started picking up in the family.

Well, the Byproducts Department actually made gas for the rest of the mill, along with coke for the Blast Furnace. The Byproducts side went into different oils, fertilizers, medicines, like Ford use to buy light oil off of us and they made jet fuel out of it. There are so many things that they made out of Byproducts with coke in it — tar, asphalt, and stuff like that. So they would bring in coal, they would superheat that, it turn into coke, which they can use to make iron. As the coke is cooking the gas is taken off, except the one end goes through a cooler and then it's liquefied. Then your tars and all your chemicals come off that, and then they go to different places. You get your ammonia sulfate for fertilizer. That was good fertilizer. It had about twenty percent some nitrogen in it, and then your different phenol, that stuff they use in T&P explosives. This area where the Byproducts was, you weren't allowed to smoke in that area. You had to take all the matches out of your pockets. You weren't even allowed to use tools made of steel, they would spark. We used brass hammers, brass chisels and stuff like that. Wasn't allowed to hook up drills. Well, they didn't have portable drills then in them days with the batteries, it was all by hand.

I was a motor inspector. Well, you're a troubleshooter. The coke has to be pushed at a certain time. They have what you call nineteen, twenty, and twenty-one hour coke. You got a battery of ovens and every half hour these ovens reverse for the gas and the heat in between the walls. So what happens after this coke is done, say, twenty hours they go in sequence. They are numbered one, eleven, twenty-one, and thirty-one, and so on, and they got a series they push. As they push the oven then they fill it up again and then they go to the next one.

They push that and then the next series might be a five so that all of them aren't cooking at one time, you know what I mean. That's the way that it's set up. The heat's turned every thirty minutes to different walls, there is a wall in between in each coke oven. It heats it so long then it reverses, and then it heats the other side of the wall and that's about it. It's dirty and filthy, I mean, you die over there. There weren't too many good days. It was filthy, it was dirty, you wouldn't believe it, when you took a shower.

I scrubbed my hair every night when I took a shower. I'd get in bed with my wife and she'd say that I stink from the Byproducts. I'd come home and would get a white towel and wipe my face and the black would come off. We worked a seven, seven, six schedule, seven daylights, seven afternoons, and seven midnights. We finished midnight like on a Thursday morning and we didn't have to go back until Tuesday. By the time you went back Tuesday the black was out of your nose, you know what I mean. It just kept coming out, you breathed it. You breathed those fumes.

Like, if we had a breakdown on a machine, we had to be there right a way, because that oven had to be pushed at a certain time and we would bust our butts. While you're on this machine, it's hot, you're near an oven, and you have to work on it. If it doesn't move you got to get it going, and our job was pretty nerve-wrecking. A lot of times we took it easy, too, but when we had a breakdown we had to be there. Now, not waiting ten or fifteen minutes, we walked up and down that road. I betcha any money we went up ten miles a day and then you put on sixteen pound tool belt and you're walking up and down the road.

The best day was when everything ran right and didn't have any breakdowns. Our job was hot in the summer and cold in the winter. It wasn't opposite because we were outside all the time. The coal handling brought the coal up to the oven and that was another story. That was filthy down there, too. That was coal dust and it was real fine, it was ground to a real fine dust. This is what they put in the ovens to do conveyer belts and we had to work on them.

It was the same thing if you would just hit a beam — all that dust fell right on you if you hit against the wall it just went all over your clothes until OSHA came in. They finally decided that this stuff was cancer causing and they gave us yellow clothes and respirators. But that wasn't, heck, I don't think that was the last ten years I worked. We wasn't allowed to take our clothes home and wash them anymore. They gave us these yellow clothes, just a jacket and pants, but the rest were our own.

So we had to actually keep our clothes there at work. You wouldn't take them home to wash, they were too dirty. No, my wife washed in two washing machines because the black dirt would go in the pumps and screw them up. Then when I come home I put my clothes on the porch and I shook them out because there were cockroaches in the lockers and everything else and you didn't want to get in your house. The work was steady, that was the good part of it. I never collected an unemployment check in all them years and that's about it. I think what I would like to forget is the way I was treated when I first went there.

When I first went there, Italians weren't . . . you know what I mean. I'm talking about discrimination. I happened to walk into the boss's office and I just transferred over from the Open Hearth, because I was going to quit the mill. I was going to go back into the Navy and stay in the Navy, and I had lost my card and I knew this boss from West Aliquippa. I grew up with his kid and this one guy grabbed it off of me and he said, "oh we'll put him on the crane." I said who the hell are you? He said, "I'm the motor inspector." Then I said you mind your own goddamn business and what war was you in? He said "I wasn't in a war." I said then you shut up, you know.

Ever since that time they made it kind of rough for me. Like you started off in them days at F - rate then from A, B, C, D, E, and that was four cents an hour raise as a craftsman. I was an electrician. I took it up in high school. That was what I was interested in, and then every step of

the way you had to go to the union to fight it. I had to fight it every step
of the way.

Even when I served in Korea and come home. I left as a Class B
electrician and the boss took my military leave of absence. I said did
anybody else make A rate while I was gone, and a couple of guys did. I
said well, I want that rate. He said, 'no you can't get it'. So I pulled out
that military leave of absence. I said you better read this, I said, if I go
to the VA you're in trouble. So they made me that rate and kept this
younger guy off and I always had to fight for daylight. I was on three
turns for about twenty-nine years and I never seen my family. When
you're working midnight you don't want to talk to nobody, you already
had a miserable night. You're trying to sleep. For seven days my wife
didn't talk to me, when I was on the midnight shift. That was from
midnight to eight.

There were advancement opportunities when they started the
apprentice program. Then they started lightening up when LTV took
over, but then things were already going down hill. I think everybody
knew that, that it wasn't going too last much longer. You know how
people do, they sell stuff off and that was it. People knew, they had a
sense of it coming. Not only that, when LTV come in they brought a lot
of young people in there that knew nothing about the mill, nothing about
the operations. They were just out of college and they hired them for say
1,300 to 1,400 a month. They asked me to go on salary and I wouldn't
take it, I said I was making more money at that time.

They asked me if I wanted to be a subforeman. First they asked me
if I wanted to go on salary because I was one of the oldest electricians
there. I said no, I've been working here to long with the men. But if you
want me to be a subforeman, I'll be a subforeman under one condition —
I work steady daylight and I don't get called out at night, and they said
that that was okay, so I did that.

A couple of times my phone rang at night and I wouldn't answer it.
To be honest I wasn't going out there after all those years I put in. I
figured it was time for me to take it easy, and then I think I stayed on as
subforeman for three years. Then they asked me to go on salary and I
wouldn't take it.

Yeah, subforeman you stayed hourly, all you did was they gave you
ten cents an hour more or something, but you drove in the mill and that
was the only advantage. Then I just gave it up and I said I don't want
anymore, I'm going back on my job cause I knew I was the oldest electrician. I'd never have to go out on jobs, I'd be a shop man. I would

work in the shop, repairing motors and doing shop things, and that's what I did until the day I retired.

The last day I was in the hospital. I forget what operation I had, but Melvin called me and he said, "guess what we got your kicker you don't have to go back in the mill." I said you mean I don't have to go back in that mill no more. He said, "yeah", he said, "you've been *Porky*" he called me. He

Byproducts coke battery and car

said, "I'm telling you, you don't have go to back. You get out of the hospital, sign up, you'll be retired," and I never went back after that. I just went back to get my tools, what I had, my personal things that I left, and never went back. That was it.

Well, I'll tell you what, I would tell a young kid to go get an education. That's all I can say. Never go in the mill because you're a slave, you got to do what they tell you and that's it. If it wasn't for the union, I think we would've all been fired a long time ago. I would've been fired ten times. That's what, you can't impress on these young kids how good a union was. They can't believe what we went through, when you came out of the mill. You were so full of dirt you had to have a drink, you drank.

I would drink for an hour and a half before I went home. You just kept coughing up and coughing up and I quit smoking. I quit smoking, oh hell, it's almost thirty years now since I quit smoking. Because of the mill, cause I breathed enough in the mill without smoking, and then I had a bypass put in my leg and I told the doctor. He said that I couldn't keep smoking and I said I quit smoking twenty-eight years ago. He said it doesn't make any difference the damage was done, I said what about all the smoke I breathed in the mill, and he said that that didn't help either.

I don't know, like I said there weren't to many good days, but I worked with a good bunch of guys, you know what I mean. Everybody worked together and if you needed help you had it. Today kids go out and get jobs, they don't know nobody, you know what I mean. There was more family. The mill brought people together, cause everybody was after the same thing, you know, making a living . . . the American dream. Before, like I say, it was just like a family. If you were building a house or if you asked a couple of your buddies to come out, they used to come. Like myself, a lot of my buddies came and helped me lay the brick. I did the same thing, built my house in 1953 and I called my buddies. I say, hey, come on help me, and they were up there the next morning.

Byproducts coke oven battery

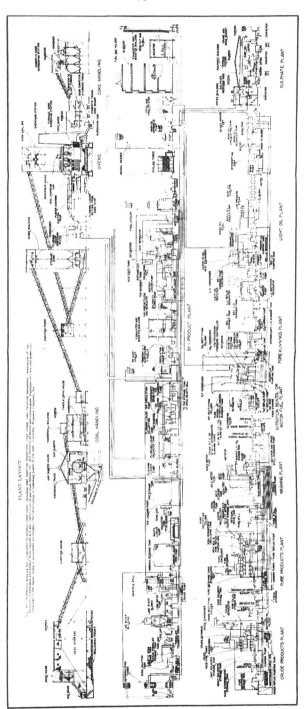

Byproducts design schematic (Reference 3)

Ed Reed and union committeemen

We met in a small, neatly kept row house with memorabilia from a lifetime kept in keepsake cupboards and bookshelves — Elvis, big band albums, and trophies. He was a towering man, six foot four inches, still agile from years of work. As a younger man, he towered above his colleagues, neat bow tie, clean cut, proudly serving as a union steward. With pride we explored newspaper clippings and magazines concerning the rise and fall of J&L Steel.

My name is Ed Reed and I am seventy years old. I was born in Lock Four, P.A. or North Charleroi, now. I was hired on March 13, 1947 and worked thirty-six years. I retired early, 1983.

I worked at J&L in Aliquippa, in the Byproducts Department. Yeah, I was in the maintenance for thirty years. First of all, in the Byproducts Department, they hoisted coal from the river and put it in the mixing bin, then the mixing bins went to the battery bin, and the battery bins, they charged the oven with it. The battery bins cooked the coal. I can't remember how many hours now off hand. From there the gases and liquids went to the Byproducts and the Byproducts — I forget what you call it now — that's where they took the different products: carbons, ammonia, sulfate. The Byproducts of coke production from coal were sold. They shipped the tar out on the river. Some other parts of that went to the Benzol Lab and they made different products. Off hand I couldn't tell what all it was. Yeah, the Benzol was shipped by river, too, and different parts of that. Some of it was shipped out by rail cars, I guess.

I started there as a millwright helper, oil greaser. From there I went on turns with the millwright helper. I worked on turns for I don't know how many years. Maybe five years. The millwrights repaired all machinery, all break-down repairs, conveyer belt, and all the machinery that runs the conveyer belt, plus the machinery. The pushing machinery that pushed the coke through the ovens to the charging car, then to the quenching car, and they went from the quenching cars to the quenching stations where they cooled the coke with water. Then they brought it back and dumped it on the wharf. From the wharf it went to the screening station where they screened different sizes of coke. So the quenching station cooled the coke, and put the fire out. So they put water on it, put the fire out. Then the screening stations separated it by size. Separated it and then they shipped it out. They sold some, shipped some to different places like Cleveland and where have you — all over, and that's pretty much a break down, as far as I can remember.

Well, we come down and start on the conveyer belt when they become worn or torn, we repaired them, we spliced them with Flexco plates, I think they called them. We set-up. We be all set-up with two come-alongs holding the belt together. Then we would proceed to cut the belt, square the belt off at the ends, and drill holes in the belt to put the Flexco plates in. I can't recall what other blades we used now, and after we got it all together, spliced up, squared up, and everything spliced and all, we put it down and the job was down.

When we repaired a chute in a coal bin, in other words any plate, any worn plate could have to be exchanged based on uneven wear. We replaced them and the same way when the coal was hoisted off the river. We had a bin there that we had to repair, plus the crushing of the coal, and then all the pusher. We had to. Sometimes a rim would crack or they had problems with the rim. We repaired that through weld and braces.

We also repaired bearings on the wheels, we jacked the coke pusher up, and replaced the bearings. We had a level and bar on the pusher that leveled the coal that was being charged in the oven. We had to repair that sometimes. It got too hot and it would bend and we would have to straighten it out. Plus repaired cables on the level and bar would become worn over a period of time.

The biggest day that I remember was when we had the big snow. I can't remember what date it was. But the snow was so deep. We were shutting down the batteries at the time to go on natural gas and we had the crane out. The batteries were fired by their own gas, byproduct gas, then we converted to natural gas fire. We converted to natural gas when we weren't making coke to keep it on a standby. But we got the crane out and running, we went to the storage area where we kept all pipes for connecting up natural gas. We had to dig. We had to wait for the crane to get through all the snow. Well, the snow was pretty close to three foot deep. This was in the 50's.

That was the thing that stands out most to me for being outstanding working in the millwright gang. That sticks in my mind pretty good, cause we worked quite a few hours, I know. I came home in a taxi. The next day I walked back to work. In other words, I worked about sixteen hours, I guess, something like that. The big snow, that was the biggest one we ever had.

I guess one of the worst days that I remember would be repairing the conveyer belt on the number two wharf, when it was all torn to pieces. We had to replace the whole belt, and it was raining at the time. We used the crane to pull the belt out plus we had to use manpower, pulling by rope and all that.

That was one of the hardest jobs, I believe. The rest of them were, well, jobs we had to do to keep the place running like that. One of the, I don't know how you say it, but one of the stinkiest jobs that we had ever had was working in the ammonia salt tower, where we hauled the salt out.

You get in there and you could hardly breathe. That was repairing a belt too in there. It was pretty hard to breathe, but if you had a cold you went in there and when you come out you don't have no more cold. It was all drained out from the ammonia sulfate. That has to be one of the stinkiest jobs. Outside of working in the Benzol Plant itself, working with the pipefitters at that time. That had an awful smell to it, too.

No, I was never in anything that much, the only time that I feel that my health might've been affected was when I worked on top of the battery when we changed what we called the stand pipe. We had to work through all that gas and smoke but when I had enough I had enough and got out. I feel that I took care of myself like that. That would be one of the hottest jobs that we ever had. When we had to change the stand pipe or change the charging hole casting. We did that with the bricklayer, or changed the tie rods that held the batteries together. That has to be one of the hottest jobs.

We had a good work gang. We all worked as a team. Naturally, you had somebody that slacked off somewhere, but we had a good team. It depends on the job. We had anywhere from a millwright and a helper or maybe three or four millwrights and a couple helpers on a job, depending on the size of the job.

I can't say that I would really like to forget about the mill because it gave me my livelihood. Raised three sons from it, built my home, own my own home. I retired from there, one thing that I would like to mention is that I went to Aliquippa high school, I didn't graduate. I went to the service in 1942 and when I was with Aliquippa high school I said I would never work at J&L, so here I was. I put thirty-six years in there.

Well, I felt that I didn't like the atmosphere and the way I went about getting a job I talked to one of my old neighbors who was a boss. William Hall was one of my old neighbors and he was the assistant superintendent, and I talked to him about going to work in the Byproducts and he said they got a job steady daylight. I said that sounds good, so I started in a day or two. I don't think there was any effect of the work on my family. My wife was working at the time, and after we had our kids I told her that she was a housekeeper now. No more work and that's all I can say, I don't think it affected her in any way. I worked, like I said, for about five years on turns — that was while I was in the millwright. After I got out of the millwrights I got in operations when they built the new coke battery, and I became a preheat operator. Then I worked three turns then. That's one of the things that made me retire early. I couldn't handle those midnight shifts anymore, midnight to eight was to much.

Well the last day was kind of funny to me, but me and my boss, Charlie Pike, talked it over and I was supposed to work daylight for my last couple of days. But we had problems scheduling. So I agreed to go back to my regular schedule midnight to eight turn to finish up my time. So my last day then I worked midnight.

Pretty long, the last day I worked midnight was pretty long, but I got a memento out of it. I took a time card and had it punched and signed by the policeman when I left the gate. I brought that home as a memento, yeah. Well, I can't really, I don't think that I could really talk to young people today because I'm from the old school, through experience in the service with the United States Coast Guard. Through that experience I learned a lot and how to do work and how to obey orders, which I don't think young people today could do it. They couldn't hack it. I'll probably think of something later on.

My work as a committeeman was tiring. We took all grievances that any member had and wrote them up and presented them to J&L to go through the grievance procedure. This is a beautiful picture here when we first started J&L, when we first started up that new place. I was a preheat operator, see the long sideburns. I think J&L was more for the man, the working men themselves. We had our differences out being a committeeman of the union. But I thought it was a good company to work for, they treated a man well over the years, and as far as LTV taking over, I don't think they were as much for the man as they were for the profit. I think that affected the men in a big way, but I don't think they performed their work well over the years through the time I spent there as an LTV employee.

Well, I think it could've been changed a little bit more by them showing more of an interest in their employees, by coming around a little more and getting together with the employees like J&L use to do over years. Well, they used to get together with different departments and talk to the men. That stopped, I think, when LTV came around.

Coal tar dump

Jim Byrd

He came under his own power, walking in the snow on a bitter February day, shoveling the driveway of his elderly compatriot who could drive. He hooked more spry than his ninety years. A powerful black man still, close-cut gray hair almost balding completely with green tinted old style glass frames, still in charge and independent. "Don't hurry me son," he said.

My name is James Byrd and I'm from Aliquippa. You got to give me time to say it, I'll be ninety years old the 12th of June. I was born in Georgia, LaGrange, Georgia. There were two kind of Georgia in those days. That was out in the country, but the city was LaGrange.

I come to Aliquippa in 1937, and was hired in the mill. I worked thirty-three years, eleven months and nineteen days. Working in the mill in 1937 was like hell.

J&L was just like Georgia Chain Gang. You heard about Chain Gang? When I come here, we had to work and get ahead of your schedule pushing coke. In the Coke Plant, we pushed this coke to get ahead of your schedule to get a chance to eat. There was no scheduled break or lunch. No, no, there were no lunch breaks. The only time you could eat was if you got ahead of the schedule.

Yeah, one day we were sitting down eating and the big boss come up, Bill Cole come up and the fellows starting running. I said what they runnin' for, I said that's the damnedest thing I ever saw, that's worse than anything I saw. So, well, I was there a long time. I tamed Bill Cole, by myself. I got them fellows together with me.

You see when I first started there, I didn't have a chance to go to school where I was at, in Georgia. Brought the Lord with me all the way. No, well, school what it was, yeah. Down there I got to fourth grade, just to read and write.

Then I came up here and started working in Byproducts in 1937. 1937, yeah, I worked down eight months and then they laid me off for eleven months. Went back to work and they never laid me off no more. I worked in the mill. I didn't drink, I wasn't a bad fellow.

Well, it was hard. Well, see, they made that coke in the oven. You was runnin' all the machines and everything in the Byproducts. I started out coke guide. I started from coke guide then to hot car. Yeah, you pushed the coke out. You ain't never seen the Coke Plant? Well, then you know what I'm talkin' about then. Uh, runnin' the hot cart, that cart, the coke. That's how they cool the coke, they quench it then bring it back to the wharf, and then put it right through the roll. Put on a roll and then they put it inside the roll. Then the wharfman they send it up to . . . they loaded it into the cars. Okay, so you had the hot car that went to the quenching station, and once the coke was cooled, it went to the loading car and then went throughout the mill.

Management was pretty bad in the 30's and 40's. The black folk was segregated. We were segregated in the Byproducts. They had white man job and the black man job. He was the heater and he was the helper. I went in the Heating Room, and they didn't want me to go in there. I told them I wanted to learn the job. They said no you ain't allowed to learn the job, but when I quit they were doin' all the jobs. The union come around and helped us all.

Well, they fired me and the union got me back on the job. I was doing two jobs and I told them I couldn't do both of them. And they told me if I don't do it, I'm goin' to do this. Then I asked them for some help on the job. When I asked them for the help, he said they weren't goin' to do it. I said well if you don't do it, I have to go home. He said if you go home, I'm goin' to fire you. I said well, I'm goin' to have to go home. I can't do that job. Well, I went home that Sunday evening and Mike Ketchum, me and Mike went down there and got me back on the job.

Well, I was doin' a hard job, it was hot. It wasn't so hot it wouldn't have been hard. And I looked to see that loop goin' up side, that hot door come right down and loading the mud. See, at first off them old batteries was hard work. The new batteries, they were self sealed doors on the new batteries. On the old batteries you had to seal the doors with clay.

Well, see, they had a heat, they pushed or discharged the coke from the furnace fifty oven in the new battery. I can't even remember now how many pushed on an old battery for eight hours. They had an exact schedule but I can't remember exactly a time when the schedule . . . it was like ten, twenty. No, the new battery they pushed fifty oven in eight hours, and they want to schedule so many minutes an hour. Sometimes you'd get behind and then they push right together in a hurry. Then they get ahead, they pushed one, then they wait, then they push another one. But if they get behind they push right behind one another.

Well, when I first come here, it was a little town. See back up on that housing plan, they wouldn't allow me up there, me and my people weren't allowed up there on Plan 6. No we weren't allowed up there. Put you in jail, yeah. This here was a bad town.

Georgia they had Klu Klux Klan was down there. Do you know anything about Klu Klux Klan? Well, I've seen them in action. I've seen them in action. They killed a fellow, dragged him down the street. Dreadful. No good reason and then they worked the people. They worked and made good crops and the Klu Klux Klan would take everything from them.

When the union come, well I'm goin' to tell you, when I started, you know what the wages was? Take a guess. Sixty-two and a half cents an hour. When I retired you know what I was makin'? Sixty two dollars and fifty cents a day. That's some kind of change and that came from the union. The union, well, we got raises — not during the war, we didn't get no raise. After the war then we got raises, I told them. I told a fellow down there, let me tell you, it took me three years to get one vacation, one week vacation. Then you know how many weeks I got of vacation? I got it three times, I got thirteen weeks of vacation a year, paid vacation. Thirteen weeks, yeah. My last thirteen weeks, I put nine on vacation and then I retired then.

Well, after we got the union it made it good. Let me see, what year it was, from '60 on. It became good in '60, from '60 onto I retired. It was good and wages were good, and then the people respected you. Roy

Die, he was cussing out one of the boys. When he was off the service, he was talking about this and that. It got under my skin, I told him off. Roy Die, J&L manager at that time. He was talking about how bad the colored people was, you know. It got under my skin, and I told him we had white, went to the service, they treat them nice when they got back. Got black they had to go save your skin but, see, no skin when they come back here, God damn it, you got a dagger and stick it in his back. When I told him that, he got red all over. That was something to remember, when I told him that. That stopped all the talk. You know what the guy got to be after that? He was my friend, he got to be my friend, yeah, Roy Die. Not a bad fellow.

No, he was the manager, one of the head managers, him and a fellow named John Gear. John Gear had a brother in the south and I went back home. He wasn't doin' too well and I got down there. I asked him, I said, John, I know you come up here the first of the year, when they were hiring. Now how could I get him here? We were going to lose a whole lot of time, I said, he is working down there. He got a good job. He give me the application blank I sent to him. He said when he sends it back to you, you bring it to me. I did that, and when my broke brother come, I got him. He left down there on a Sunday and he got here on a Sunday night and Monday morning he went. We took him down to work and he got a job and he went to work Tuesday morning.

Well, John Gear, when I went in the employment office, oh man, they know me and all, I told him, come on, I took him all around there. They all commence to hollering, about the long line, they got to hollering about it. Then after a certain time, my own brother — after he got to workin', got to makin' money, he got to be drunkin', — he would lay-off. I told him, listen, he was always tellin' me about who is talking about this and that. I said, let me tell you one thing, "now, a man who got the job for you told me to talk to you, and you better do right."

Well, he mess around there, he messed up, so he got in trouble. Some kind of trouble there and they put him in jail, made him pay a fine. On Saturday and Sunday he had to go down and lay in jail, and at the end of Sunday they let him out so he could stay on his job. We helped him out like that.

I had the best woman, Carrie B. Byrd, in the world. Lord, Lord, Lord, she was a wonderful woman. We ended up married in 1932, sixty years. She died in 1992. Well, her daddy . . . I didn't have nobody to help me, she had a pretty good education and she helped me. So when I worked, I'll tell you how I got ahead in Aliquippa, I was playing the numbers to buy me a house. That didn't work. In '43, I quit playing numbers and quit tradin'' on the credit. In '47, I had enough money to buy a house and pay cash for it. I'm living in that house now.

I have two kids. Two good, I educated both of them. They're wonderful children, they are both school teachers. They work and one of them, the younger one retired. She got . . . her health was bad. The little one is still working, she got three girls and her husband is assistant superintendent in Pittsburgh school district.

Well, they treated the colored pretty good in the last days. I wasn't there when they shut down. Well, see, I retired in 1970. They were good, from 1965 to '70. They were good, I made good money. Yeah, full pension, both ways. I got pension from social security and J&L.

Well, I can tell them young folks they had to work and to go back and forth on jobs. My buddy, me and him, his boss got mad with him. Said he was crazy, they were goin' to fire him. He didn't want to leave, he went there for twenty-three years, never laid off a day, and wasn't off a day. They told him. His boss said he was crazy, go ahead, we won't fire you, but he went down there to the head psychiatrist. He told him, said, looky here, this man got a perfect record, I don't see that red flag nowhere. Said what's wrong with him. He said "I just don't like him". He said, "Well, if you don't want work with him, there's the door, if you don't want to work him."

Well, J&L was all right by me. The first vacation I got, some of guys quit drawing their pay. When it was time for them to go back to work, he got hungry, but I draw my pay everyday. But when I left, the union got us help.

Coke "Beehive" ovens

The quench: cool 3,000°F processed coke

ALQUIPPA WORKS
COKE PLANT

Description of Facilities

Make of Ovens	Koppers
Number of Batteries	4
Number of Ovens	352
Type of Ovens	3 Batteries - Becker-Underjet
	1 Battery - Becker-Guntype

Production Data

Coke - 1,935,000 N.T. Annual Capacity
By-Products - Tar, Ammonium Sulphate, Gas, Light Oil, Benzol, Toluol, Xylol, Crude Solvent, Residue, Pyridine, Sodium Phenolate

Operating Statistics

	A-1, A-3, A-4	A-2
Tons of Coal Charged per Oven	17.25	17.40
Tons of Coke Produce per Oven	12.20	12.20
Average Hours Coking Time	17.00	19.00

Equipment Data

		A-1, A-3, A-4	A-2
Ovens	- Length	40'-5"	40'-8"
	- Width - Pusher Side	15-1/4"	15"
	- Width - Coke Side	18-3/4"	17"
	- Height - Floor to Crown	13'	14'
	- Capacity - Cubic Feet	693	705

Reference 2

Chapter 4

ℰℭ

Steelworks

He was the group historian, chronicling the past — old wage tables, photographs and employment cards. However, this bespectacled pensive man helped with the future — making recommendations for health care programs, and political agendas for the other retirees.

My name is Michael Teleha and am forty twice years of age. I was born in Barnesborough, Pennsylvania and worked forty-seven years, forty-seven years and three months officially. I was hired in 1931, maybe in November, November 3, 1931, I remember the exact date I was hired. My wage was about a hundred and sixty dollars at the time for two weeks. My hourly wage was thirty-three cents an hour for a ten hour day. Three dollars and thirty cents a day, my first pay envelope was six dollars and sixty cents.

This was my job classification system. They went from laborer and so on, and each job was worth a certain amount of money. In other words, it depends on how much you did, and what effect it had on the production, was how these rates came up. This rate was established in 1947. It runs from job class 0, which made ninety-six cents an hour, to job class thirty which made a dollar ninety-eight cents an hour, double.

Most job class 30 was like in the open hearth department, the closest they had to was 24. The first helper with the guy who controlled the BOF after the old Open Hearth Furnace. The Steel Works actually performed the process of melting iron ore and making steel ingots. The steel ingots went into the Blooming Mill and were rolled into pipe — seamless pipe, welded tube

Michael Teleha

pipe, tin-plate, or wire mill. I was in the steel making process, so they brought in raw iron and we melted it down in the hot furnace to processed steel ingots. Well, let me put it this way, they brought over iron ore and they put it through the Bessemer process, they blew all the impurities out of it, and then they took it down to the Open Hearth department. They put it in a Open Hearth furnace and processed it for maybe five or six hours by adding certain elements to it and to affect the amount of carbon in the steel. Its different properties were achieved by adding elements such as copper, or whatever it is, to make steel.

In other words, every grade of steel had to be processed under a certain different method. If we made stainless steel we made it with one process, and anything else is a different process. We went from high grade steel, medium grade steel, down to a low grade steel for the mill.

In other words, some of the stuff went to the Blooming Mill, in the Blooming Mill they rolled out blooms, they rolled out slabs, they rolled out plates, and they rolled out big sheets of steel for the tin mill. So this is the process in the Open Hearth Furnace.

In our steel works the Basic Oxygen Process was the same, but it wasn't what they did there. They used a certain amount of raw tin or raw material like scrap iron, scrap tin, and so they melted hot iron from the Blast Furnace and they run that oxygen probe and blew out — blew in or blew out — all the impurities. We used to work, we used to make steel in the Open Hearth department maybe five, six, seven hours, or whatever it is, and the basic oxygen we were making a heat of steel — 100 or 200 ton of steel — about every thirty minutes. It was much faster.

First, they built the Basic Oxygen Furnace then they built the Standcaster. They pulled up the other side and then they made certain blooms. In other words, they were six-by-six, four-by-four, or two-by-six inch sizes. There were different size blooms of steel that went through there, and most of it went through either in the Seamless tube or the Rod mill or the Welded tube area. This is where the steel came out of the basic oxygen furnace, not including the Strandcaster. So the Strandcaster was coupled with the Basic Oxygen Furnace. Well, they made the steel and they also made the steel in the open, in the basic oxygen furnace, and transferred it — took it upstairs, and run it through these molds.

My first job down there was laborer. Well, I did everything. They assigned you to different departments, to clean-up or to do this. Well, you did almost anything and everything. You piled bricks, you shoveled coal, you did this, you did almost anything and everything that had to be done in making the steel. In other words, if your a laborer you swept up floors, you did unloading stuff, you loaded different parts or different elements, and you went back-and-forth to different things. This is the first job that I had. Everybody starts as a laborer, and then they work their way up.

Well, I would say there were quite a number of people who knew the right people in the employment office, who would end up with better jobs in certain places. Especially if you were a relative of somebody, like a foreman down the mill and stuff like that. You always ended up with a better job than laborer.

Well, how did the people feel about that? Had no choice. We're talking about the height of the Depression. There wasn't no choice, they said, hey Mike, you want to work in labor gang. Yeah. So I went to laborer to Open Hearth laborer to general laborer to riggers. I worked in a Rigger department. I never worked in the tin mill, but I worked in every other department in the mill as a laborer.

The Rigger department. Actually that's what they were doing at that time — building up buildings. Now this was a part time job, in other words, I went down to the employment office looking for work when things got tough, so they assigned me to the Rigger department and Blooming Mill and assigned me to Tin Plate. In other words, they assigned me to different departments in the mill where they needed a laborer. I was almost in every department of the mill, went all over the place.

The best day I remember in the mill is when I got the job. I was hired in 1931 and I worked for awhile, maybe four or five months, on-and-off, one day, two days, a week. Then we would give you one day a month to pay your insurance. In other words, your insurance was a dollar twenty-five cents, you got a thousand dollar insurance plan, so you had to work one day so they could take out money to pay for your insurance. This was the way of life in the steel mill at that time. So you went and did this and that was that.

One day to work and pay for our insurance and, well, when I got rehired in 1931, '33, that's when the war efforts went into effect with the National Recovery Administration (NRA). After we won the fight, then our rate and everything went up. We went from a ten hour day to an eight hour day, so they hired an awful lot of people as laborers.

The National Recovery Act, that was one of Roosevelt's rehabilitation programs. This was when the government forced the National Recovery Act and they went to an eight hour day from ten or twelve hour days. On top of that, the Japanese started to buy an awful lot of steel, a lot of scrap. They were buying all the scrap that was available from the Open Hearth at J&L. They were shipping all this out and hiring more and more people. So actually, if you cut the length of the day you could hire more people. Well, sure, when you went from a ten hour day to an eight hour day, so you went from two crews to three crews.

The worst thing I remember was when I worked in the tin plate department a laborer died. At that time we had a job shoveling coal in through the gas producers. We burnt coal to make gas to heat the tin plate. In other words, they heated it up, and I worked there three days with a big scoop shovel. I ended up with huge blisters on both hands.

Took the next couple of days off and then, when I went back to work a couple of days later, my foreman told me — my boss at that time — he said, "You know mister, there are a hundred people outside looking for your job, if you don't want to work say so".

I showed him my hands were blistered from one end to the other. That didn't make too much difference with him. He said what he was going to do and that was it.

Outside of the fact, I was applying for a different job, when I was a laborer and I went in the Blooming Mill. I was about eighteen or nineteen years old at that time, and found out they were hiring crane operators, so I talked to the foreman. I can't remember what his name is, it's not really important. So I went to the foreman and I told him I heard you're hiring cranemen, and he said yeah, we are hiring cranemen, do you think you could run a crane? I said, yeah, I'm pretty sure I can run anything if I have to, if I get a couple hours break in. He said, well, what is your name? I said Michael Teleha. He said Michael Teleha, what nationality are you? I said, I'm a Slav or Slavic. He said, "Oh, we have no openings for cranemen."

I still remember the fact, and this was no joke. That's when I told him I was a Slovak, and he looked at me and he said we have no openings at this time. And now I knew that they had, somebody told me, that they were hiring crane operators in the Blooming Mill. There were good jobs in the Blooming Mill at that time. You made more than thirty-three cents a hour at that time, my wage was. An eight-hour day my wage came to three dollars and seventy-six cents, but that was under the National Recovery Act. Yeah, you never forget things like that. I can remember well during the Depression during 1930-31, I used to go down to the employment office everyday in the morning and everyday in the afternoon. At eight in the morning you would wait around and the foreman would say, okay fellows, nothing going on today, go home. So you would come back at four in the afternoon hoping that something would open up then. And we kept doing this, day-by-day-by-day, and every once in awhile if you were lucky enough, or if the guy liked you, he'd give you a job for a week someplace. But what it actually ended up was you ended up outside looking in. The guy used to chase me out of the employment area twice a day for a month, this went on for years, from 1931 until 1933.

So when I got a job in 1933, I didn't know how to react. I was a laborer in every department of the mill, I worked in the blast furnace, and one of the things I could remember: I remember when I got a job in the blast furnace for a week, five days, ten hours a day. I would work from seven at night to five in the morning. I could still remember when I got down into the blast furnace at that time, I was only a little over eighteen years old, I'm pretty sure. So I'm over there all eyes and mouth, wide-opened mouth, looking around everything wondering what the hell is really going on.

So, I asked one of those old Serbian men, what do you call it. I know he was Serbian, I could understand his language more than anything else. So I said to him hey mister, and he said, "what do you want kids?"

I said "What time is that furnace going to tap, what time is it going to make iron"?

He looked at me and said, "hey kids, you go to school"?

I said, "yeah, I go to school.",

He said, "what's the matter, book no told you".

I said, "no, book no told me." This is the way of life in the Open Hearth.

I can remember this down in the Open Hearth department when I was laboring there. You would come out at eight in the morning. You would come out at four in the afternoon, and then you would come out at midnight looking for a job. If they needed somebody, they hired, they put somebody to work. Now, everybody that wanted to work had to have a shovel. In other words they would come over like soldiers and they have their shovel on their shoulders, and if the foreman needed a couple of guys to work, he picked the guys with shovels and the guys who didn't have any shovel would go home. They hid their shovels and somebody would watch where they hid their shovel, and then they stole them. So next time they had shovels looking for a job.

Now this was a way of life in the early 30's. If you wanted a job, they wouldn't buy them. The foreman would hang himself before he would put an order in for a handful of shovels. So if the general labor gang would come in occasionally with a bunch of shovels and picks and stuff like that. Then everybody stole a shovel from the labor gang. So they had someplace to go to work. Like I say, this was a way of life in the early 30's in the steel works.

This is where I spent most of my life from the time I had six years in the old Tin Mill. That was the old Hot Mill, I used to run a scrapper. And this was six years — 1932 until 1938 — I was on and off. The day I got married in 1930 — no, I got married in 1938, and got laid off the following week.

So I was working, I got laid off for a week then they called me in for a week's work. I signed up for unemployment compensation. Now unemployment compensation was a big deal at that time. All of fifteen dollars a week. So you had fifteen dollars and one week you might work two or three days of a regular job, or the following was to sign up for workmen compensation, or you signed up for two weeks of compensation and you might be lucky enough to work two weeks. Like I said, this was the way of life in the Aliquippa Works in the early 30's. I worked with a bunch of guys. We hunted together, we fished together, we did everything together. The guys names were Al Goonis and Jimmy Grossi. Yeah, I mean we worked together, we socialized together, we did everything together. We even chased different girls together. There wasn't too much of that social life in the Steel Works area during that time. It was hard work.

Everybody was looking for work, everybody had their own little system, everyone had their own little fiefdom. There wasn't too much social activities in the mill among the employees. In other words, hard work. We worked with black people, Serbs, Croats, Slovaks, almost any kind of person or nationality was in the Open Hearth department.

They come in and they were coming through, one of things I would say when this occurred was shut down the coal mine. When they shut down the coal mines a bunch of coal miners were brought in as laborers cause they were a little short on laborers. So they brought in a bunch of coal miners and they didn't stay too long 'cause everytime one of those cranes went overhead, there would be rumbling, and those guys would duck ready to run. They said they never had this in the coal mines, we can't stay here, we're leaving. So this is what it is.

This is about the only funny thing worth repeating. The coalminers they just didn't like the idea of them cranes going overhead rumbling and noisy and all that stuff. They just couldn't take it and most of them, quite a number of them, went back to the coal mines and some of them stayed. An awful lot of them traveled in between California, PA and Aliquippa, and some traveled everyday and others rented a room someplace.

Well, I remember when I started work in late 1939-40, we had one event when they were pouring steel and one of the ingots blew up and killed the steel pourer. He was burnt and died a couple of weeks, a week later, and a couple of guys got burnt pretty bad, but outside of that. Well, at that time in the 30's, they had to send them all the way to

Pittsburgh to Southside Hospital. They had their own medical center down here in Aliquippa, and the main office. A doctor would come in, but they didn't have any facilities to take care of burn patients or bad breaks or anything like that. So if they had any kind of special break or something else, they would send everybody to Southside Hospital in Pittsburgh. When they built this new hospital up here, why, then they sent everybody there. Well, the company donated the land. Admiral Morell was the president of the J&L organization and he promised to put so much money in for every dollar that the steelworkers collected. So we put money in, too.

So there I was, a union committeeman at that time. I went to three different departments. I was a steel works representative in the steel works. I don't know how many years we did this, but we went to different — three different — money raising events and collected money for the Aliquippa Hospital. This is one of the things the fundraising for the hospital included. Well, everybody, they wanted everybody to donate a certain amount of money per hour. Our work, whatever it was — one dollar or two dollars — or whatever at that time — and we you know but so you signed up and they took this money out of you pay you know it was only a couple of dollars two, three, or four something. Well, we built this hospital and it was very worth it. It was exceptional, and over the years I don't know what happened, I don't know if it was administration or whatever happens to hospitals, but it really went down the drain to a certain extent. But it's coming back. It was probably worth it. No doubt about it. Not only for the steelworkers, but also for everybody else in this area.

The effect of the mill on my family, well, yeah, I got married in 1938 and I started working in 1931. Well, my wife went along for the ride. What else could she do? Right, was she worried about it? Yeah I did have a dangerous job, there was no doubt about that. You'd hear about someone getting hurt or something like that and she would worry. I have four children, I have two boys and two girls, all born in the late 30's and early 40's. The two boys did work in the mill. They worked part-time during their summer vacation while they were out of school, because both of my boys went to college. And both my daughters went to college, too, and my one son graduated from Clarion College in Clarion, PA. My son, John, graduated from Geneva College and my daughter, Dorothy, went to Slippery Rock College for a couple of years. Well, she was there for two years and then she got married, and my younger, Joyce, went to College up here in Pittsburgh.

If I could change something in the mill, I would keep it working. I don't know how, but I would make every effort to keep it working. What LTV did to the Aliquippa Works shouldn't have been done to a dog. They aborted the entire process, they shot it all to hell. One time the Open Hearth at the Aliquippa Works was the most productive and financially set plant in the country, and all the money we made in Aliquippa then built-up Hennepin, the Cleveland Works, and Otis Steel was up. Until what LTV did, between him and LTV and this Tom Graham, he sold J&L out to LTV. We went down hill. I don't know when the last time you were in Aliquippa, but look around. In 1982, when J&L was cutting back, they originally screwed up the Rod Mill and they went to the Welded Tube Mill, and slowly but surely the entire works sometime in '82.

December 31st was the day I retired. Now, I know that it was my last day and I worked like any other normal day. When I got done I took all my stuff that was mine that could be taken out, I took that and out I went, and that was it. No, nothing special, just another day at work. Yeah, just another day at work. In other words, nobody gave me any special dispensation. All the guys said have a nice time Mike, and so on and so forth, and that was it.

What can you tell the younger people? Everyone has their own method and their own way to live, and what they want to do and how they want to do it, and they don't like anybody else telling them what could you do or what you should do.

Aliquippa golden jubilee parade

Open Hearth Furnace sample

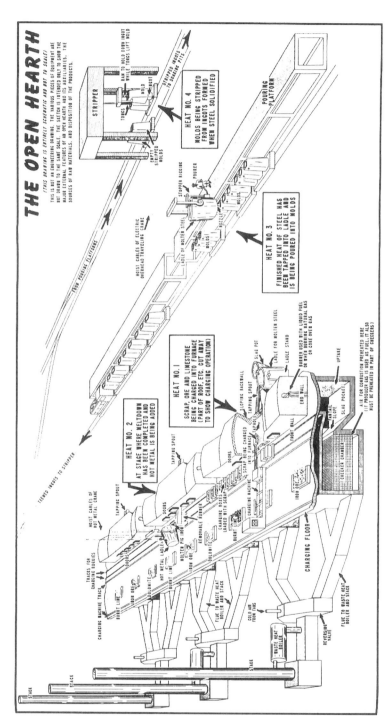

Open Hearth Furnace schematic (Reference 3)

Patrick Darroch

He was a quiet, stern man, huge in proportion with almost gray-white side swept hair. The bitterness and hurt was still present ten years after the shut down for good work unrewarded. He used to be a man of significance — craneman, union position, but with unusual, vivid clarity he described the shut down along sterile but cruel financial terms.

My name is Patrick A. Darroch. I'm fifty-two years old — will be fifty-two in a couple of days — and I was born in Hopewell.

I was hired in 1964, April, in the Tin Mill, and worked twenty-three and a half years. I shut the plant down in 1985, and was narrowed to the last eight to leave in October of 1985. My cut off date was October 1st.

My wage was less than twenty dollars a day when I started. The hourly rate was around two dollars and something. I didn't take that home, it was less than twenty dollars a day for eight hours work without shift differential. When I retired, while there were different parities as a swingman and craneman, I bounced around with incentives and everything. Probably around thirteen dollars an hour.

I finished up in the Steelworks, BOF, and Strandcaster, and they called it the BOF Steelworks. We had both facilities, the Strandcaster and continuous caster. The Steelworks made different heats and grades for the other departments. We had the strandcast, which increased productivity by two hundred percent when it came into existence, before I got there. The continuous strandcaster took molten steel and made finished products-bars or billets, eliminating the blooming process where ingots needed to be reheated to form these products. We had to make the heats in the BOF, where we made the hot metal. The BOF is a Basic Oxygen Furnace, where we used to pour hot molten iron in. Then we would pour the ingredients or additives, and then have an injection of high pressure oxygen, and turned it into steel. Different grades of steel usually take up to 3,000 degrees Fahrenheit temperature for strandcast heat, but they ran all grades of steel there — high carbon, low carbon. High carbon steel is used for your better machinery, auto models, aircrafts, some aircraft components, mainly auto industries like the drive lines, the axles, and axle housing.

Low carbon steel would be used for generic products, what they would run into auto bodies. Stuff for refrigerators and Tin Mill products, depending on what grades they finished with. So then, when the Basic Oxygen Furnace would have a heat of molten steel, you would pour that into ingots by the old method.

In 1985, probably one out of every four heats would go to the Strandcast, 'cause it was a six strand continuous caster. It would make billets which, in turn, they would be used for guardrails and trailer beams — housing beams for your house trailers. Main products was the housing trailers, mobile trailers, for the frame. Also your I-beams for your safety guards or barriers for the highways. Then other stuff went through the Rolling Mill.

We came out with the bottom four there, which was a cleaner steel. It was more refined and decreased the waste on the strandcast. The yield was usually around ninety to ninety-three percent, which was maximizing all your steel efforts. It also cut down on the cost per ton per yield. Like I said earlier, when they first came out with the strandcast, the 14 Inch Mill would make structural stuff also — flats and angle irons. They increased their productivity two hundred percent with the strandcast.

Then, at the end, we made rounds and something with the German machine. Nobody, the Germans themselves with their own machines, couldn't even make a seven inch round — the first thing we started up on after four months with the change over or the shut down to retool. They turned around and started up with a seven inch round and something the Germans couldn't do was a seven inch round with their own machine, and we did that there.

The BOF technology, it was going to slabs. Your ingots would go into the Blooming Mill, and from there it would go to different grades, all depending on the grades and what the product was used for. We would pour it into the slabs or ingots and the continuous strandcaster lets you to get to a little more finished product. Semi-finished product also eliminated a lot of our jobs. It probably eliminated about nine or ten processes. It was a semi-finished product is why they utilized it, and that was modern.

The job was hard, it was dangerous, as anything in the Steelworks was. Hot molten steel had no mercy for nobody. It was hard job but it was well-compensated too, and the guys had a decent living.

In the Steelworks I labored for two weeks. Done anything from shoveling limestone that broke off the conveyers, cleaning tracks, to up in the overhead precipitator, which is all your dust and everything that was overflowing and everything. But it was general labor and then after two weeks I went on overhead crane, which I ran before. I really didn't want to get back onto overhead cranes, but it ended up a necessity. That was where the money was and that's where I went.

Out of the twenty three and a half years, I ran crane in the Tin Mill about three years, and Steelworks I ran it for thirteen years. I was in three different departments. I started in the Tin Mill, went into the Apprenticeship Program. They busted the Apprenticeship Program up, and I went back to the Tin Mill. Then I bid up to the Steelworks, and that's where I finished my time up.

Crane operator was a lot of responsibility. It was dealing with a gantry with a man on it. Nobody could be underneath you. You would be in the scrap yards moving cabs and it was hard. You couldn't get off the job. You would see the guys at the washhouse when you start and then at the end of the turn. But we had radio communications throughout the whole BOF.

Foreman would call and say like on a Scrap crane, "We need sixty thousand pounds of scrap," and we would calculate our scale. Then set-up and try to throw the right mixture in for that heat which was a strandcast heat. Then we go up to maybe a 100,000 or 150,000 pounds depending on what heats they were calculating for. They had a two million dollar computer that, at the end, they shipped out elsewhere. Then our melting foremen would calculate all this through that computer, and then call us with what he needed for the scrap, and that was one process.

Another process was on the Charging cranes, which I would fill-in for because seniority-wise, you needed thirty to forty years on higher paying jobs. They were direct bonuses on a twenty-four hour basis, but Scrap crane would only be paid eight points and Teeming crane would be paid sixteen points. That's where you would team the heats and that's where they pick the ladles up and pour the steel over top of the molds, like Mike Teleha did. He was a steel pourer.

It was dangerous work there. You never knew what, once steel got on top of water it would blow. You could put water on top of steel and you didn't worry about it to cool down, but the vice-versa. We had incidents down there, with the molds blowing up and guys getting burnt, and you had to always be on your toes. It's twenty-four hour function and it was mass production, heavy machinery. You had to be aware of stuff. I ran all of the cranes in the BOF, Scrap crane, Teeming crane. These cranes were rated to lift some sort of weight. On the scrap crane we had a forty ton capacity, the magnet was one of the biggest magnets on the East Coast. It just fit inside a railroad car and it was about eight foot in diameter, and had one for each one of the two cranes in that bay. Then you had two charging cranes that would pick up, their rate of capacity was 275 tons and you ran multiple trolleys. There were two trolleys and three sets of hoists, but they tied off the one hoist.

Our ladle capacity was 250 tons cause they increased the size of them and then the strandcast heat they could only take up less. It had to be less than 190 ton, 'cause of emergency purposes, 'cause the ladle would burn through. They had a E-ladle up there and that was the capacity it was able to handle. But as far as the cranes over there and the hoists went, they were rated at 275 ton.

My best experience: probably getting in the Apprenticeship Program, Boilermakers, that was in 1969. So just getting in was helpful. That was almost assured at that time that it was a lifetime job. Never in the prior history was an apprenticeship program ever laid off. They would cut back to four days, but he was never laid off. It was almost sure of a guarantee job for life.

Definitely, I was surprised when the change took place, it hurt me. I just bought a farm, had thirty and a half acres. I was unloading the farm tractor and I got a call not to report out, even though I was on schedule as apprentice. That year was 1971.

My worst day, probably the day outside contractor jumped in the ladle that I was supposed to be picking up in a hoist for the strandcast. The guy was a pipefitter I believe, an outside firm. He ran and jumped in the ladle and burned up. The issue that was there that my heat was coming.

Yeah, the guy did it deliberately. Also, the way guys got killed in there. It was demoralizing. It happened often, more often than . . . more often than the company wanted to take for. If it was an outside person, like that guy they didn't even count on the safety standards. We had outside painters fall. As a matter of fact, another bad day was I saw a young guy, an ironworker, fall off one of the cranes that he was erecting and he broke his neck, and there was nothing we could do for him. There were some bad incidents down there.

There was guy who got called out, one of motor inspectors got called out on an off day. He was retiring that week and he got crushed in between the crane and the girder, and his helper was lucky that he didn't get caught also. The helper got up through the side of the crane and got on top of the bridge of the crane, and Johnny didn't make it. It's just fate that you're so close. Yet another foreman, had a couple foremen that got killed down there. One got electrocuted, had 400,000 volts hit the back of his neck. He was a pretty good guy. Another temporary foreman fell off the crane and he died.

Another bad day was a different guy got killed in there. One guy was an outside construction man who was working on a crane, he fell. I watched him fall through the air and he landed and broke his neck. Another part-time foreman fell off the crane. I wasn't in the plant at that time. A guy jumped in the ladle, he was an outside person and it was the heat that I was waiting for, and I didn't know why it was coming up to the strandcast. It was tapped out from the BOF furnace, and there was a guy who jumped in the ladle and burned up. That was a deliberate thing, he was . . . rumors were he was attracted to fire.

Unfortunately on top of the ladle they would throw rare earth on top of it to try to contain the heat because on the variations. Once you drop below 2915 degrees Fahrenheit on most of the grades, that would be 2,915 degrees Fahrenheit, it would start to harden up on you, and you couldn't pour it upstairs so you have to take it down to the teeming area.

My most memorable event was winning the committeeman's job for the steelworks. I was on lay-off status, was out in street and I beat the incumbent. I thought he would give me super seniority and go back in, but it turned out I was going in to get my reinstatement physical the same day that three and a half hour meeting of the indefinite idling of the Aliquippa Works was taking place . The nurse in the afternoon, whenever I got my physical, she said, your pulse is awful high. And I said, if you would've sat through that what I went through in the morning you would've been worse off, heart attack or something.

Uncertainties were what I would like to forget most. Very much uncertainties when we were on lay-offs. You never knew when you were going to be recalled. We were always on a roller coaster. I never worked five years steady, we always jumped. It was a seasonal thing for some of those things, that's why I went to the Tin Mill and that's why I bid into the Apprenticeship Program, which I thought I had a lifetime job. Then I went up to the steelworks and it was good money up there, but it became a seasonal thing, too.

Mikey Dolnack, I labored for two weeks in the BOF, when I first went up there. They needed cranemen very badly and I ran a crane in the Tin Mill, so I went up on the job without being properly broken into the working conditions of that crane, and what the job duties were. Mikey put up with me and tolerated me and helped me along. He was an outstanding man. Unfortunately he's dead today. He was a young man when he died.

Well, my two daughters, I never seen them. Was always on shifts. The two daughters are the oldest, there is one twenty-five and one twenty-three, and my boy was born in '76. One of the things with all the lay-offs, I was able to spend time with my son, but then you had no money to do anything, either. You between a rock and a hard place.

If I could change something, it probably would be to put up some type of a fight to stop the acquisition of Republic Steel. Their pension funds were unfunded and it drained everything that we had. LTV bought us mainly as J&L. When J&L was there our pension fund was very sound, and after LTV got us they kind of more or less raped the assets is what they were after. And then they picked up Republic with no pension funding. Top management from Republic somehow or another came in at the top end of it. Seemed after that all jobs left Pennsylvania, they're in Ohio. They changed their corporate headquarters from Pittsburgh to Cleveland. I think if we would've made a more stronger, sound fight back in the older days — whenever they wanted to acquire with approval of Congress Republic Steel — maybe we could of stopped them.

Yeah, their compensation and wages are the same. They are closer for the shipping, for they are right on the lake for shipping iron ore and stuff, which cuts down on some of the costs. But their compensation and insurances were a big factor. The workman's compensation factor. Yeah, I had a chance in 1969 to go in the apprenticeship program in Cleveland. I didn't take because the cost of living was so high at that time I was living in a co-op, but I regret that fact, too. I should've went up there

Steelworks towering over West Aliquippa

Steelworks floor — personnel and ladle

Steel cast

because I would still be working today as a boilermaker/ironworker is what they combined the job up there for. I think the combination of Youngstown Sheet and Tube and Republic Steel was an endeavor to switch into the Ohio environment. Yeah, they acquired more properties and duplication of services. But the funny part of it was that we in Aliquippa Works use to break records in our Blast Furnace and our BOF. It was the same design and same equipment that Cleveland had, and Cleveland couldn't do the same things we did. As a matter of fact, we taught them things. We would go from a high grade heat down to a high sulfur heat, and then go back up to a low sulfur. That's just what we did there, our . . . we had guys with college educations and we had working men there that worked with each other. Hot molten steel doesn't respect nobody — management or salary — it will burn anybody. So you had to work together.

My last day was in September of 1985. There were eight of us working there. I knew it was the end because I was done. I looked around and unfortunately I seen "Vote Republican Reagan". Some of the guys had been put on different parts of the mill and are still there. I don't miss the quietness of once massive steel producing machinery and noise to the total silence. Another thing was the position of the LTV salary guys. They came out with the phrase "Liar, Thieves, and Vultures" because they treated their salary people. Like dirt. They had nothing for them, it was terrible.

"Liar, Thieves, and Vultures" the crew of foreman came up with the idea because they didn't tell them the truth, and they didn't deal well. So the new Republic Steel was probably the initiative behind it. The mill provided a good quality of life, our standard of living along with the union benefits. Today's life for my eighteen year old son, what he has to do to acquire something in the future, and it's not unless you have a quality education.

The Steelworks

David Henderson

You could just tell that he was fiercely proud of the work he had done in the Steelworks. A compact, solidly built man, light skinned, balding with a roughish goatee currently working as a security guard. But when he told tales of the mill, the work they had done, he was proud with a glowing sense of accomplishment, as he wove intricate stories of men and steel with a rapid, almost staccato southern accent.

My name is David Henderson and I live in Aliquippa. I was born in Union South Carolina.

The latter part of 1955, I was hired. I can't get that completely, but I was working labor in the General Labor pool, and you know about what they were making back at that time.

I ain't workin' in no mill. They pushed me out. Well, as a matter of fact, they gave me a retirement. They asked me if I would retire and they would give me a good amount a month, and would I take the retirement? I said, well, sure I will. That was the Basic Oxygen Furnace (BOF) out on the island, the Steel Works. I took early retirement in 1982.

The Basic Oxygen Furnace I first started over, well, in the Labor Gang there. Then I moved up to the stock house. That's like material, you know, I mean, magnesium, lime, and stones and stuff go into the heat. Then I moved from there to the service floor, like a service man. Wherever they need to go in the heat. But some of the different ingredients they put in from the service area. When I was a serviceman I had to take care of the lance, so the lance is clean, you know. When you blow, and the lance comes up with all that, like, bark you blow into the heat, I have to clean that down there. That was the servicemans job.

So you had steel and then you took an oxygen lance and blew oxygen into it to cause a hotter heat, and then that came out with some contaminants on it and you took that off. Well, in the labor gang, we mostly cleaned around. We were up in the — this is not the correct name — but we called it the pipes, on the front of the furnace. We had to clean all that that got stuck in the side from the coal. We cleaned a lot of that. It was pretty rough.

Well, I did that between six to nine months, and that's when I went into the stockhouse. Like, I was material man, and I don't know what it was called. Cars come in loaded with lime. Then we had to unload what you call that, hopper or fifth floor. I can't think of that floor up on top, where we send that lime. Come up on like a belt and it would come up over the top and go down to the particular bins — but what was that floor called, I can't think of the name of it. It was the 5th floor of the BOF. Yeah, I came down to the — no wait, um, sorry, I moved from the stockhouse to serviceman. Then I moved from serviceman to head stocker, what they call it, head stocker.

Then my next move was down to third man on the furnace. During that particular time they had three men — first helper, second helper, and third helper. I took temperatures then, the read would come out in the pulpit, you know. Like when they blow the heat, they want to send temperatures from the heat, what we had that was running. Then the read would come out in the pool pit, you know, to operators in the mill. The operators and melters would see it all.

Let me see, a good day, let's see, well, I had some pretty good days. Even a couple of times I moved up to operator. That, I thought, was just about a perfect day, because they take their time with me, which I hadn't been properly broke in. But I did what the melter had told me to do. I blew some heats. They were surprised that I blow that well myself. 'Cause, as a matter of fact, I blew one, a special heat that they told me that Indiana Harbor Plant couldn't blow.

The guy told me, I hate to tell you this because you look like you're doing such a good job but, whenever Indiana Harbor learns this you're going to lose your job. So I said, what you mean. He said they are here to learn how to blow this heat and then shut this mill down. He said, yes, we're going to do that so I'm just telling you that you enjoy it so much. This is the biggest job that you'll move up to. I said, in the Basic Oxygen that's when you move up to operator. I say you can't go no higher than that.

Well, I'll tell ya, the worst day I remember was when I was in the stockhouse, cause sometimes it would get so cold down there, man. Lot of time when I first started working there we didn't have what they call shanty, we had something to go back under, like, uh, I don't know what you would call it. But it had four posts on each side and it had something like a bagging sheet over it. You go back in cold. They had a coat jack outside, but how that was, sometimes I would get the coldest down there.

But I would tough it out, cause sometimes when me and my buddy . . . when we got cold we would spell each other off the line. We had these big tankers and Tim would say, why don't you go get warm and then you come back over and then I'll go and get warm. Then we could keep up see because we would just set the material up. I think that's one of the worst days of my working in J&L out on that island, cause a couple of days my feet got so cold I could hardly . . . I didn't walk. I came from out there and went over in the mixer holes and sort of stood there for awhile.

'Cause now when you say my best days, coming back to some of the best days was in the mixer holes 'cause I poured iron, I poured iron. Oh my goodness, that was before I moved up to second man on the furnace. I poured iron, you talk about pouring iron, man, I think I poured about twenty heats that day and they had the engineer to put. He took it apart what they wanted, then he would tell me how much iron to pour, and then I kept my feet on that trap. But I enjoyed doing that and they knew I knew the job.

What they would do a lot of times if they needed somebody, they would ask me to work over. Other than that it wasn't a hard job in the mixer hole, it was just, uh, you had to be on the alert. But I say from being that operator, that was temporary, for three days, that was typical. Because sometime, they were going real fast and they would tell the first helper to go charge that furnace. Okay I got it charged. A lot of times they would tell me well go ahead and blow the heat. The melter would be ready, they wanted a heat analysis from the mixer hole. Those were my best days.

My worst days were in the stockhouse, I had some rough days down in the stockhouse in the winter, my feet got so cold one day they froze. I knew it because I couldn't walk when I left there, but then they built the shanty and they put heat inside the shanty. You know, and then they have these cars coming in the same way and you would open them up and send the lime or whatever material you wanted, and it got pretty rough there, too. But a couple of times I was on the stone car because those things were freezing, and we would be down there with a torch trying to melt it out. I say I didn't know the stones would be frozen, you got to let the gas heat go all way up through there. They were some pretty tough days, but like I said my best days was down in the mixer hole to the move up to being that operator.

No I didn't have any problems because I'll tell you why, I tried to work with everybody. I tried to work equally with everybody, and some people said, well, he's a white guy, he's a black guy. I didn't see no white man doin' my job, I didn't see no black man doin' my job. I told a lot of them that I'm just the man doin' my job. If he wouldn't work with me, well then, he didn't have work. One or two times I had a couple of guys that weren't working. I said, now wait a minute, you're pulling just as much of the load that I'm pulling.

I'll never forget when they hired the ladies. I had a lady out there working with me and I was trying to help her, but she was pretty rough and she said I can't get that, and I said, well, I'll get it for you. She said, wait a minute, I don't want you to do my work. I said no, I'm just trying to help. She said, I can pull it.

I'll tell ya, sometimes she would be struggling with that big old cable on that car and pull it. I told her, I'm not going to try and be ballsy but, I said, when you get to something too heavy that you can't do, you just ask David. I'm goin' to tell you this, I don't think I'm wrong but, you see, women weren't made for this rough work.

I think sometimes, though, some of them do it for the proof. But I told her, I see how you're straining. That cable is big and that thing was heavy, man, and you know we would be pulling the cars. I told her just call me, but I said if you don't and you go ahead and put it on, I said, okay. But just call me if you get to anything that you can't do, I said, call David.

I'm the man, you're a woman. She said, well, David they pay the ladies the same amount as the men. I said, yeah, you're getting the same amount of pay, and she said, well, then I do equal work. I said, well, if

you can't do it, I said a man would be less than a man to make you try
and do something that you can't do. I'm not saying you can't. I'm
saying I see how you're straining trying to hold a cable this big with this
big old iron hook, you know. I said, don't do yourself like that.

I said don't do it like that, I'm a gentleman, I'm not no violent man
or nothing. I said it's not going to hurt me one bit, to share a little bit of
my strength. I said the woman was made for the lesser weight, and the
man was made for the greater weight. She said, all right, all right dear.
I said, whenever you work with me down here in the stockhouse, don't
you be afraid. I said, you tell David. Say David, I can't get that, and I'll
gladly come and do it. Because, I said, if there's something I'm doing
and I couldn't do it so well and I needed your help, I believe you would
help me. I tell her, I see some of you women working up to being a
mixerman or operator. That's the easiest work, but you have to go up
the line.

The mixerman just calls for material and that's all. 'Cause sometimes
if the guy doesn't understand figures to well — which I have a little
trouble catching on — the helper would keep up with it for him, and all
that and the operator, well, all you had to do was look at what the mill
was bringing down and put that ingredient in the heat. Those were my
best days. I would say the mixerman, and then when I moved up to
operator, because in the particular line that was as high up as you went,
when you got moved up to operator.

Let's see, a co-worker . . . I would say, well, one of them has gone
on that was Jackson, he was good co-worker. Roselle Boyd and Jim
Sims from Sewickley. He was one of the guys who gave me a talking to
when I messed up down there. He wasn't . . . he was sort of a friend in
his own way, and it seemed to me that he didn't want you to ask him too
many questions. But when I started working outside, he'd just go right
ahead and he would pour a lot of iron, and he would say do you want to
pour it?

I said, how do you get that there, so I thought, okay, I know what to
do. I got my paper since I was slow with figures, and I would watch
when he would pull, and I would write that down and the amount he had
to get from the other pot. So I wasn't bein' rude, and me and him made
a joke out of it. But he was that kind of fellow. I think his name was
Steve. I can't think of the last name but we always called him Steve.

At the end of the day he had a smart way, he would look at my paper
so he said where did you get that from, and I said, uh, I think I'm writing

this down 'cause I took my papers home. When I was studying I said, now, you a good man, but you wouldn't let me see what you have. He said don't do like that Dave. I said, look now, you see you got some iron outta that other part, and that's going to balance your figures. He and I made a big laugh out of it, and he told me I'm just that kind of a guy. But the next time I told him I needed some more breaking in. He said, well, you had a week. I said, now, look, that's not enough for this guy.

Sims, he was operating at the time and he had trouble. He said, David, I got disgusted when I poured some iron up. I pulled the pot down to the floor and I'll tell you what, I got nervous, and instead of bringing the pot down and letting the iron drain out of it, some of it went down on the scale.

We had a big thing out of it, but I had to get most of that iron out from under there. I admit it was kind of hard but, I said, it won't happen anymore. I told him what happen and I said I was goin' to quit, and Sims said, David, don't let them push you out of that job. I said, why. He said, Dave, I know you know, he said, I'm not as fast in figures as you. He said, if I can get it then you can get it. And I thanked him for that. Sometimes, he and I used to call it up on a good bit.

Then he was on to a new trick nothing to do with religion, he sanctified you know. Sometime they set themselves off kinda on the side now.

He and I would talk a lot, and I told him that's he is a good man, so I's come up sometime when I be working. I told him I would kind of be watching him, and then a couple of times after blowing the heat, I would watch him. When I was working first helper, he got to the point that sometimes he would take charge of the front, and that kept me out of that operation. When I got around, when I blew a couple of those specimens, he said, well, how did you do it David? I said, well, I'm through watching, I'm always eager to learn about something that I think would help me. He said, David, "sure nough now, it's no secret you could blow this, I bet." I said, I did what the melter told me to do. He said, are you satisfied with that? Move to it.

I said, all right, and I went back in and found some of the old guys papers. It seems with this particular heat, I said, I watched everything he did — the way the guy moves the line. He said "don't tell my secret." It's no secret I did it at one time. He told me to make sure I didn't overblow, so I said a minute and a half before it was time to pull it, I figured they had added some scrap.

It was going to take quite long, and I said, come here and push it. He said, well, tell me you didn't blow it. I said, well so what, do what the man over you told you to do. He said, I don't over blow it. I said, now if I had went to four times then I would overblown it then, but then you couldn't do nothing. But see, if I underblow, they can always come back and heat another minute or minute and half. The third one I blew, I recall, the guy said, so David, you want to blow the heat? I said, yeah, so the guy said David blows it.

Well, I did it when he said so. They said, well, let's hit it, and the melter said, no, don't hit it, he said. Why, I said. He said, because if you hit it you'll knock the carbon out. Wait till we see the analysis and if it didn't come out the way it is supposed to. Then you go ahead and do it. But if you do that now, I'm goin' to put it on you as releasing the carbon and it come in perfect. And a lot of old operators said, I can't see how you can blow heat like that. You are not an operator. I said, I was willing to watch, and we made a big laugh out of all that.

I said, don't you notice how I sort of be noticing. So the guy said, you know what he does, when you come in this pulpit he'd be watching and noticing. I said, man, I'd be so hope'n and praying for my day to come, but it didn't come. I said they slid the board out from under. But he said, I'll tell you what, sir, I think you would've made a better operator than you do a mixerman. He said, I told the guy you were just about a perfect mixerman.

I said, no, I wasn't perfect. I made a mistake one time, is what I said. The guy told me, pour a certain amount of hot metal. I said, but he didn't say how much. I know when they make a strandcast heat you pour around 260. So my mind was set on 280 for the regular heat, I poured the 280, and took a temperature. So the melter who calculates the heats called down and he said, why you put that much iron in. I said, well, usually I put in a certain amount.

He said, well, man, if it goes over 280, what am I goin' do? I said, well, you didn't say how much to put in. He said, you already put 250 in. I said, all right go ahead and be hard on me. Haven't I been a pretty good guy. He said, you've been a perfect guy, Dave, but it just so happens until now. I said, can't you calculate it and tell them and tell the craneman to hold back some of the iron. He said, what am I going to do on my next heat. I said, you work out something. You missed one night, and you come down here?

I said, well, then don't be so hard on me. Just make this heat come out, so it just so happens he calculated it for 275. He said, I'm goin' to gamble on this and I'm goin' to go ahead and pour the 280 in there. I think I can make that up in slag, and it did it come out perfect. And the guy said, who do you have down in that mixer hole? My friend said, they got a good mixerman down there. You tell him what you want he is goin' to put it in that ladle. If you calculate them heats right, I've been watching this. He said, they're waiting on you, Dave, when you come in there, 'cause they know you know how to work those scales. The guy told me, well, I'll tell you what, I'm not braggin' on that he's one of the best mixerman that we have. I told him, well, I thank you for that, I appreciate that, that comes from watching pretty close. He said, well, what are you working towards now, Dave. I said, I'm working towards operator.

Well this story isn't exactly a great story, but I always chuckle about it. Okay, my buddy that I was working with, he get in that furnace that metal goin' to come out of there. Just so happens that I was off that particular day, so he said, David, they blew the furnace up and half went the other way on. It blew that whole wall out on the side. What would you have done if you were here? I said, you know what I would of done. That's all I told him. I just made a big laugh out of it, I said, you know exactly what I would have done.

I don't know that what I am sayin' I did tell him. I might said I'd have hurried to the men's room. Me and him made a great big old laugh out of it, and I said, you know, I kind of like working out here. But I still don't like this. That's why I didn't come up through the pit, I didn't like how those guys sometimes lost the heat. Oh, all that iron, I said, no, I don't want to be over here and I got out from over there and went to the other side where it was a little safer.

I have me and my wife, we weren't lookin' for children. Well, she adjusted herself to the shifts real good when I decided to work on swing turn. 'Cause when we first got married, I was in the General Labor Gang and I was steady daylight all the time, unless one of the older furnaces went down. Sometimes, they would put me on four to midnight, you know, until that would be finished with that. But she adjust herself to it real good.

Well, I'll tell you what, if I had the chance and go back and start it up again, I would be operating in full blast, because my times over there now. Secondly, I think I would want it to be just like it was because, you see, they was in full bloom and was goin' real good. Understand the only thing, a couple of time it had the oxygen line froze up for a while. But I think if it came, now they could change that, you know. I think we were getting our oxygen from West Virginia or somewhere it came in.

But a couple of times the line froze and when they froze up, then there
was no oxygen coming through. But other than that I think things should
be the way it was.

Well, I really don't know why things went bad. One fellow told me
that he thought the reason why we wasn't there, the way things should
have been was no longer. He said, Crucible Steel down here in Midland.
They have electric furnaces. Well, see, we have these oxygen. He said,
I believe that if they had not been long out on that island . . . and then
they shut it down like that. He told me and that's the only thing I can
think of. He said, well, see, they don't be bothering with a lot of that
scrap and stuff. They go by electric. But I never work around an
electric furnace.

But that is the only thing I can see. That must've been what happened
because, honestly, when they told me they were goin' to shut it down, I
said, I don't believe that and when a fellow told me from Indiana Harbor.
He said, man, you are all doin' some powerful work. I said, you don't
know what you're talkin' about. He said, yes I do, the bottom is goin' to
drop out and everything. He said, as soon as that island mill is running,
you're all goin' to be out of a job. He said, but I like the way, you know,
I like the way you blow that heat, perfectly. How long you been operating?
I said, I'm a green man. He said, well, I'll tell you what, sir, I bet you
make a good operator.

He asked me how much time I had and I told him, and he asked me
about my age and I said, well, they're goin' to move some of the guys.
He said, would you be willing to relocate. I told him no, because I have
made a start here. I would have to try to sell my house, then I know my
wife isn't goin' to want to relocate cause her mother and father were
living — both gone now, God rest them — but I told him no, I wouldn't
be. He said, well I think you might get something 'cause of your age and
your time and all, which I did, I got a retirement.

Last day in the mill, I can't get that down pat. But before I went on
retirement, they called me in to ask me about retiring, and the lady told
me that they would give me three months in order to accept the retirement.
I told her, well, chances are if I stay here, well, a lot of times they'll send
us home early. I said, if I stay here, I might not work. No way, she
said, would you be willing to take retirement? I said, yeah, I'll take it.
She moved it up. She said they moved it up in the contract some kinda
way, 'cause I had a perfect record all this time. I never had one day off.
It just so happened I got sick a couple of times, but I got sick during the

time when I was already home from work, because when I had this stomach flu, I knew it, and the guy told me.

I would say at least fifteen years, it might've been more. I never missed a day and I'm proud of myself. Actually I got sick one day in the mill, and I went to the office man, and I said I got to go home I'm sick. He said, well, Dave, chances are you can't take any leave, can you make it? I said, man, I got this stomach flu and I don't know whether you've had it or not, but you know what that will do to you. Oh man, I was bent over, and I said I got to throw-up. He said, go back there and just sit down. I said all right I'll do that, then I started to feel a little better. He said, don't eat nothing. Then he called down to the hospital for me, and nurse told me, when you go home don't eat nothing. See if you can drink some juice, if you can drink some juice, drink some juice, and I did that. I made it and I was off the next two days. I hate to throw-up, I hate the flu, I felt like I had rocks in my stomach.

One day I went in and the guy said, what you doin' out here? I said I'm suppose to come to work. He said I know you didn't drive. No, I had some hip boots and I put them things on and rolled down through town, and had the guy let me in the gate down there, the other gate. He said, where you goin'? I said, I got to go to work. He said, you can't go till a quarter after. So I got to go all way down to the island. He said you goin' to have to wait. I said, no, I want to go in through the snow.

Oh, that was the last time when we had that great big ol' snow. I can't recall the year, but it snowed so much. Oh, at least three or four miles. Well, you see, I had to leave home early. I walked downtown to Aliquippa, so I just walked on down to the mill. You see I had the time. I called the police department, they said it's bad and if you don't have to drive don't drive. Well, I said, how else am I suppose to go. So I said, if I'm goin' to go I'll put my hip boots on and I went walking right ahead on, 'cause that particular night I was goin' to work four to midnight. I made it. Well, he told me that I was the only man who got a perfect record, he said, that man got a perfect record.

Well, I'll tell you what, I don't have too much to tell them, but the main thing that I would tell them wouldn't exactly be about the mill. But I'd tell them to go to school and get an education where ever you go. Then you can demand a certain type of job. Now, I didn't get no further than high school, I didn't get the chance to go to college, but number one I would tell them to go to school. Now, I believe if they would ever get another mill back it would be on different basis or something like that, but I just knew after J&L ranked up there that long. I still call it J&L, and they call it LTV. I feel I should have went to college. I told somebody there ain't nothin' out there, but with J&L I got a jacket. It had J&L on the emblem, came off, but I'm goin' to have it sewn back in there. I said when LTV and the rest of them took it over, it went down the drain. I said when it was J&L, I said, it was something. It was something when LTV and all that went down drain, it went down the creek. He said, you shouldn't say that.

I say, well, they start doin' this and doin' that, I said. J&L was runnin' it and they even had a great big store downtown called Pittsburgh Mercantile, and you could go there buy clothes I called it J&L. It was a company store. I called it J&L and I had this jacket at home and I treasure it, but some of the emblem came off — some kind of ink, I can't tell. I'm goin' to get somebody like a seamstress and let them sew that back around in there, and then I'm goin' to keep it. I bought the jacket just a little bit too large when I bought it and now it's a good fit, so I told somebody I always will remember J&L. That's what I call it, J&L. Even sometimes right now. I said, I don't care what it is, I still call it J&L.

Steelworks heat

Steelworks gang

Steelworks construction

Basic Oxygen Furnace (BOF) — 70'

Steelworks floor

The continuous strandcaster

West Aliquippa Island

ALIQUIPPA WORKS
BASIC OXYGEN, OPEN HEARTH & BESSEMER

Production Data

Number of Furnances	Kind	Rated Capacity Per Heat N.T.	Annual Capacity N.T.
2	Basic Oxygen Process	81	880,000
4	Basic Open Hearth, Tilting	152	1,040,000
1	Basic Open Hearth, Stationary	152	124,000
3	Bessemer Converters	28	384,000
	Total		2,428,000

Operating Statistics

	Oxygen	Basic Bessemer	Open Hearth Duplex	Scrap
Net Tons Per hour (Tap to Tap)	115	60	41	15
Average Tons per heat	81	27	152	152
Kind of Fuel Used	Oxygen	Air	Fuel Oil - Nat Gas Coke Oven Gas	
Fuel Rate (MMBTU/WT)	-	-	2,654,950	
Oxygen Consumed (Cu.Ft./NT)	1612	-		

Equipment Data

Shop	No. of Units	Rated Capacity Per Heat (N.T.)	Hearth Area (Sq. Ft.)	Depth of Bath
Basic Oxygen	2	81		38"
Bessemer	3	28		18" to 27"
Open Hearth	1 Stationary	152	728	
Open Hearth	2 Tilting	152	704	
Open Hearth	2 Tilting	152	784	

Reference 2

Basic Oxygen Furnaces

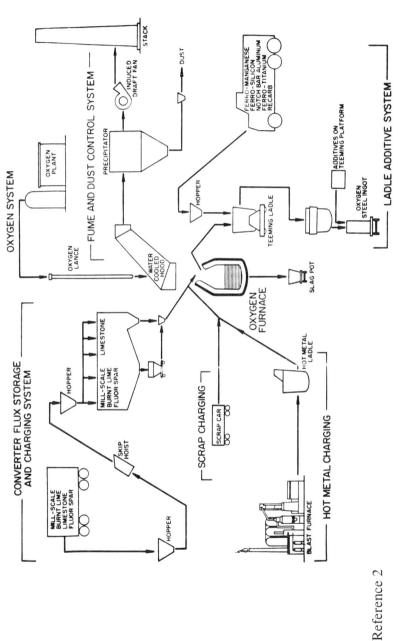

Diagram shows material flow of new basic oxygen furnace processing installation

Reference 2

Chapter 5

ℰᴏᏟᏄ

Blast Furnace

A pleasant, cheerful man related his story of working in one of the dirtiest places on earth. He smiled and his eyes twinkled as he told of the neverending task of cleaning the sinter dust an inch thick.

Kaz is the name most people know me by, but my real name is Kazimierz. I'm sixty-nine years old and worked twenty-seven years in the mill. I was hired in 1951 but don't remember that far back, and retired in 1985. I worked at J & L Steel Plant in Aliquippa, in the Blast Furnace Department — Sinter Plant.

In the Sinter Plant, I worked on mixer, which took iron ore, limestone, coal, scale, and flue dust, and combined them to make sinter, which was fed to the furnace. The scale was iron and steel particles combined with coal dust to make sinter, that made use of the scrap to allow the furnace to run again. Sinter was what we made from remnant materials to increase efficiency and make more iron. You know what I mean? You would use iron ore and all of the stuff in the furnace prior to cast, and you'd have to use less iron ore, and worked a lot faster.

There was another job there that I did before the mixer, working on the train hopper with the trestle man to bring material to make sinter. The trestle man would bring the coal, iron pellets, or scrap to make iron,

Mixer Man Kazimierz Pudyh

and then you would unload in the stock house. The stock house was an underground area that contained a subterranean railroad. The trestle car would accept a load of material from the rail car above ground and transfer it to the hopper. The big hopper mixed up everything and was drawn to the top of the furnace by pulley and cable, and that went into the Blast Furnace. That was hard work, it was outside, some on the trestle, while the stock house was thirty feet underground. I didn't like that outside work, not crazy about it. Well, it was all dirty work. The inside work was no better, there weren't many good jobs inside the Sinter Plant. Too boring. I got in the Sinter Plant, after they shut Blast Furnace Number One down, I worked in the stock house. You either get laid off or go to Sinter Plant, so I went to Sinter Plant.

Then I went to Tilter Building. From Tilter Building, all water from furnace come to us in two big tanks — the mix machine and two big tanks. I work in the continuous recycler with all the water. The continuous recycler was a machine where our water come from furnace to two tanks. They build a big pump station, then pump our water to steel tower. Then send it out from Sinter Plant to furnace. That thing, it ran two and a half years, and then they shut it down. A shame, this was a beautiful

Young Kazimierz in Sinter Plant

place and cost fifty million. They shut 'em down. It was a fifty million facility that only ran for two years. The best day I remember in the mill. No, I wasn't too happy to retire, no way, but did not have choice. They shut down, that's it. The day I was hired, oh, they were happy days. I was forty-five years old when I went to work. I worked before that, but wanted to work in Tin Mill. So, I went to work with grocery store business instead, then went back to the mill. Right, that's why I wound up in the Tin Mill.

So, 1960-61, I started the second time. I basically was in a situation where I started work, and then I stopped working to start my own business. Well, pretty good, I think. 1959, the strike year, was a problem. So times were hard and everyone needed credit. How was a store supposed to give credit? So I was unsure if when people were on strike they could pay, so I could not accept check, put sign up for cash only. I knew things were going to be bad when strike continued.

After six months on strike, that killed me. So I lost about six thousand dollars, not that bad, you know, what's his name downtown, they took him for sixty thousand dollars. Then in 1959, six thousand dollars was a lot of money.

The worst thing I remember about the mill was working for the Sinter Plant. That's my worst time, but its not that bad. It was all right, I didn't mind. Sure, I liked it and was happy working for the mill. Yeah, I guess my family liked me working in the mill. Well, I remember everything about the mill, but nothing really stands out except the fact that you could still make a good living. Even though things were bad, you were able to make a good living. Nothing was really bad in the mill, you just did your work, made a good living.

I remember lots of buddies and good friends from the mill. Dave Smith, I think about in particular. I was his helper, he was all right. That was already in recycler. On trestle, couple guys are dead who I worked with. Some people work with me, I don't even know if they are alive. The guys on the trestle . . . only one I know, he is still here. Another thing I didn't like about it, I didn't like to go to work on Sunday. Somehow, after kids grow up, I didn't mind it, working Sundays. But when the kids were small, I didn't like working on a Sunday. I remember a bad problem in Sinter Plant. Guy who I worked with had an accident with mixer. Yes, that's what I am sayin', it was sad. Everybody liked him, he was a good worker. That could have been me.

There was nothing that could have been done to prevent the shutdown. I don't think it was the worker's fault. I don't know what happened, why they shut down, they wanted to make money. They shut Aliquippa for money. They let Cleveland and Indiana work. We were always making money, ya know what I mean? I don't know, I really don't know. Why, I don't have any idea why.

The effect on Aliquippa, oh, oh, Aliquippa went down, that was it. Well, everything went down on the street. Aliquippa downtown was beautiful. Franklin Avenue, that was where the shops were, then everything went down.

The Pittsburgh Mercantile was a company store. Well, they did not have too much to do. That was the company where you would buy on credit. They take from your pay. The company store began to change, stock less merchandise, less traffic. Maybe 1960, '61, it closed.

I would tell the young people, well, something like I wish the mill will hopefully be working and don't know if they would like it, but they should work there. Yeah, I think they should still work there. The mill taught me how to do stuff, whatever I did, but for the continuous recycler,

I went to school. Yeah, we went to school for two weeks and I think I learned lots of stuff, you know, the pump worked 28,000 gallons per minute for the furnace. That's a big pump rate. That was forty to forty-eight inch pipe. We had to make sure it was working. I had to make sure the water was moving to cool the system. Right, that was my job to make sure they got enough water.

My last day in the mill, I worked. Clean up, clean up, clean up, I wanted to leave it spotless, no dust just like I found it. I did not want to leave it a mess before shutdown, but I wanted to make sure the place was clean. The mop, the broom, it was a nice place, then I left. And we shut 'em down, Blast Furnace Number Two.

Blast Furnace Aliquippa Works 1-5

"Tapping the hole" Cast House floor

Ore pit and yard blast furnace (November 20, 1907)

Blast Furnace

Hopper car with Blast Furnace charge

Stockhouse Larry car

Slag alley — skip car

The Mill

Blast Furnace stockhouse gang — Bill Stephens afternoon foreman

Stockhouse control room

Cast house floor

Submarine iron ladle

He had every aspect of ironmaking permanently inscribed in his memory, as the details were related piece by piece, operation by operation. Their days were not marked by hours but the number of casts and shifts.

My name is Francis Yakich and I'm from Center Township. I'm fifty-five years old and was born in Pittsburgh. Actually twenty-four years and nine or ten months was the time I worked, after being hired in January 17, 1962. I was making about two dollars and seventy cents or two dollars and ninety cents, something like that, an hour. When I retired in 1987, it was the end of the year, I was making roughly, I think, about twelve or thirteen dollars an hour.

The Blast Furnace, actually, they melted down the . . . well, they made the iron from the limestone and iron ore and a combination of materials. They melted it in the furnace and then casted into iron ladles, which went to the Basic Oxygen Furnace. So materials came through the ore yard, which was unloaded from railroad cars by cranes. Material was loaded into the trestle cars in the Stockhouse, which was twenty feet underground and would go to the top of the hoist, get dumped in the furnace. Then you would pour iron out of the bottom into ingots or ladles. Well the ladles would be hauled down to the Basic Oxygen, (at that point in time it was the Open Hearth Furnace in the earlier years, until the Basic Oxygen was developed) and they would turn it into steel.

Yeah, just a couple of years before that, I think, I really don't know what year the Basic Oxygen came on, but I think it was in the 70's. So basically, then, they would take the ladles. The iron was hot and they would pour that into the steel vessel. Yeah, they had, well, they were submarine ladles was what they called the iron ladles. They were shaped like a submarine, and they were lined with brick. They would transport them down to what we called the island, which is by West Aliquippa, and dump it into the vessel down there. The submarine ladle would tilt over and you would dump all the iron into the steel ladle. Ladles were anywhere from 150 to 200 ton. Then there were the slag ladles. We called them cinder ladles. That was a byproduct that rose to the top of the iron, so the iron was heavy and sank to the bottom.

Well, the iron casting process began with a trough constructed in front of the tapping hole, which was a hole that they drilled into where the iron came out. The molten iron, like you had mentioned, the iron is heavier than the slag, slag being the byproduct. The trough was designed with a series of dams with one dam in front of it, which was lower than the dam to the side of, and like, behind it. It's hard to really explain.

The iron would come out and drop to the bottom and hold a certain level and be running into these submarine iron ladles. When the slag came out, which is lighter, it would ride on top of the iron and go to the opposite side of the furnace where the cinder ladles were. One side was iron ladles and the other side was cinder ladles, and you would fill those ladles up.

We had an area that they called the cinder yard, which had a series of gates and runners depending on the size of the furnace. There were like five or six of these cinder ladles and iron ladles. The smaller furnace having two and the bigger furnaces having four iron ladles, and the slag would run out over the top and fill these cinder ladles. They would run the same way out to the West Aliquippa area where they had a slag dump, that point out there they would dump those . . . tip those ladles over. The slag would run out and cool to the sides of the truck. At night the sky would glow orange, like it was day in the hills where they would dump the slag.

The Blast Furnace had the road, the ore yard, the slag alley, the furnaces, and the iron alley, and on the other side of that was the Boiler House. I started at the low end which was the cinder yard. The title of that job was, they called them, cinder snappers. They cleaned that area up, that side of the furnace, which was the cinder side. The runners were shoveled out, cleaned out with forks. And it was sort of like a kids play yard. It was all sand, the runners were dragged in the sand. Everytime they casted the furnace and ran that slag out you had to clean the excess that cooled off on the runners and throw it into a box, or what have you. That was then dumped into a car and hauled back out to the dump, too. After every cast you had to clean up, that was the initial job.

We started out in the early days, there were four to five casts a day. They did jump that up when we left out of there. In the late '80's we were making ten or twelve casts a day, they were going every hour-and-a-half to two hours. Yeah, they sort of . . . they use to dump those troughs and drain all the slag and iron out before. But in order to make those extra casts they would just leave that molten material in the trough before we shut down. They had different types of insulation that they tried to cover the iron with and keep it molten. Just go out through the cast again without cleaning it, which saved me time.

Yeah, I guess their theory was that the safer, actually, the more production, too. But the more safer it was, the emptier that furnace was; the more, you know, the better. You weren't keeping a big quantity in there, like if you held it for four or five hours, you had a bigger quantity of iron in there. Yeah, it was safer keeping it empty.

Well I progressed right up the line through them all. The next one is what they called the first helper. That man worked on the iron side. Basically, he'd do the same thing that the cinder snappers did, but he cleaned up the iron with the assistance of the laborers. There was only one that they called the first helper, and there were like, two or three, depending on the size of the furnace, that were second helpers.

Well, cinder snapper and second helper were the same. Then they had a keeper. The keeper was in charge of the trough and they had what they called a mud gun, which was a machine. I guess you would say with a piston in it that they put clay in to stop the hole when they drilled up and the iron was finished running. The gun was mechanical. Electric motors tilted it into position and a plunger would push the clay down to stop the molten iron from running until the next cast, and they would pull the gun out and they would have, like, a clean plug in.

The next step — I guess you would say that the keeper was one of the higher jobs and then they had what they called a stove man or stove tender. Eventually, that's what I retired out on. I was on that position, which is the top of the crew, actually. The stove tender was in charge of what we called the stoves, which are the heating chambers that melt the material in the furnace. The turbines or the boilerhouse would turn the turbines, which heats, which generates the air, which was pumped over to the furnaces. In order to supercharge that air, it would go through these stoves, which were a series of brick checkers in a big chamber. Then it was blown over down through and back out the other side, and it would be, I want say, 1500, 1800 degrees Fahrenheit. It would pick up the heat out of those bricks and then it was blown into the furnace.

Well there are four to five I think, say four stoves on a furnace, and you would heat the brick up by natural gas, or gas from the byproducts or the coke works, and gas would go over and heat up the chambers in the stoves. You would — the stoveman would — watch the gauges and charts and what have you, and determine when the stove was at its maximum heat. Then he would redirect the flow of air for one to the hotter one, and then that air would be going into the furnace. Then you would shut the other stove down that was just cooled off and reheat it, and go through that whole process. It got a little bit technical at the end. Everything was charted, and they got into the electronics and the air and oxygen levels were monitored. It was a real good sign. Before, what was originally in the 60's, it was guesswork by looking at the color of the flame. It was basically just looking at your furnace and if you see a nice flame compared to a dull one, you would heat that way. But in the late 70's and 80's they got high-tech.

There was a water tender. He was in charge of the . . . he would walk his route and check all the water. That was one the dangers of the Blast Furnace. If the water would leak into the furnace, you'd have explosions, which they did.

The water circulated around the furnace to cool it. It was in jackets, well, early on. Well, even still at the end there was a combination of brass plates that were laid in the brick work inside the hearth of the furnace from the stack down to the hearth of it. These brick were all laid around these plates, that's what we called them, and there were all water jackets that kept that brick cool when the furnace was hot from the material burning down through. The wear and tear wore that brick away, and then it would cut the ends of these plates which, if not taken care of, they would just flood the furnace.

If it didn't blow up, it would put it out, because it would be too much water with it, and the water tender sort of watched that water and was in charge of changing any ones that went bad. They also had coolers and twiers, which were the openings where the hot air blew into the furnace to keep it. You could actually see into the furnace and they were big water jackets, too, which could be coated with the molten iron inside and cause a lot of water and damage, too. Probably the keeper was the hardest job. You had to crawl down in those troughs and the temperature was hot. I mean, they went from initially starting with wooden shoes, and they got shoemakers that were out on the avenue that would tack onto your regular shoes car tire for soles. Some people liked that. I always thought they were too bulky and cumbersome, plus your feet would get so hot.

Well I didn't see them — I was on the job working at that time — but two fellow employees died eventually. They were burnt when the furnace blew out. We were working on the same furnace together; they went one way, and I went the other way. The coolers and plates flew out of the furnace from leaking water or whatever. The furnace had been down prior to this, and we had put the furnace back on and it was on probably two or three hours. Then two of the fellow employees who were water tenders . . . we had just . . . it was nine or ten in the morning. I would say nine, it was early. We had just been talking at the one back corner of the furnace and they were going to go on their way, and I was going on my way. Just as I turned to get down, I felt a concussion in the heat, and I just ran right on down the steps. But they had been going the other way, and they were right in front of it. They died a week or two later, they had burns all over ninety percent of their bodies. It's probably better.

Number Two furnace was the largest followed by Number Four and the Number One. Yeah, One, Three and Five. Let's see, Three was the first to go down. I would say, probably 1983, maybe 1984. Yeah, Four probably went down before that, but they rebuilt Four. Four was rebuilt and ready to go, and they were supposed to put it back on, and they never did. It was completely rebuilt and, in fact, they kept it heated up for a whole year. And then Five went down after that.

The dirt and the heat are what I remember most, and also what I would like to forget. Well, at first it was sort of traumatic, changing shifts were. Never did get adjusted to them. It was just an irritable midnight shift. It seemed like I got more rest on it, but I had to sleep all the time. Well, see, I got shifted around really. On my last day I was in the Labor Gang.

Well, I would say it was a good life. I mean, as far as monetarily to raise a family and that. Working conditions weren't the best in the world, but economically it was the best place to be at that point in time.

I don't know, I think they just had a predetermination to get these mills out of this area. Everybody claimed it was because of the location. They needed to get closer to the lakes or what have you. I don't know if that's true, I couldn't argue with 'em one way or the other. It was just corporate mindset to get it out of here, I mean, I thought everybody pitched together the last couple years to keep the place rolling. But it was still to no avail.

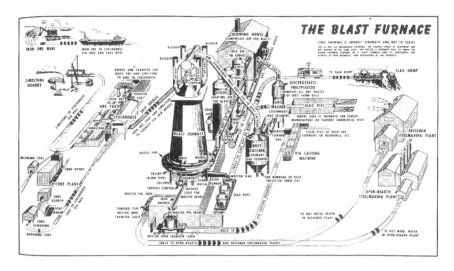

Blast Furnace schematic

Ore yard trestle

The Mill

Furnace with skip car

Blast Furnace demolition site

A-1 blast furnace; built 1909, demolished 5/11/90

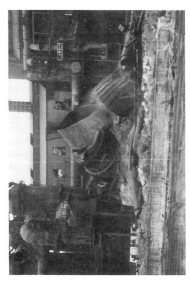

A-2 blast furnace; built 1910, demolished 5/23/90

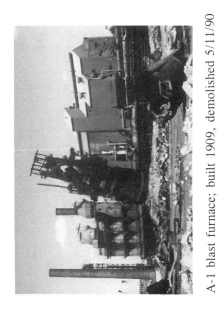

A-1 blast furnace; built 1909, demolished 5/11/90

A-2 blast furnace; built 1910, demolished 5/23/90

The Mill

A-4 blast furnace; built 1912, demolished 4/17/90

A-5 blast furnace; built 1919, demolished 4/24/90

A-3 blast furnace; built 1910, demolished 5/15/90

A-4 blast furnace; built 1912, demolished 4/17/90

ALIQUIPPA WORKS BLAST FURNACES

Production Data

Furnace	Rated Daily Capacity - N.T.	Annual Capacity N.T.
A - 1	1,214	442,950
A - 2	1,213	442,950
A - 3	1,214	442,950
A - 4	1,073	391,750
Total	5,726	2,090,000

Equipment Data

Furnaces

Furnance	Height	Hearth Diameter	Working Volume of Stack-Cu.Ft.	Bosh Diameter	Date Installed	Date Last Relined
A - 1	88'-3"	28'-6"	33,700	30'-0"	1909	1949
A - 2	88'-3"	28'-6"	33,700	30'-0"	1910	1957
A - 3	88'-3"	28'-6"	33,700	30'-0"	1910	1948
A - 4	88'-3"	27'-0"	32,575	29'-6"	1912	1951
A - 5	88'-3"	26'-6"	31,520	28'-9"	1919	1960

Stoves

Furnance	Number of Stoves	Heating Area Square Ft.	Fuel
A - 1	4	584,400	Blast Fce. Gas
A - 2	4	465,300	Blast Fce. Gas
A - 3	4	541,600	Blast Fce. Gas
A - 4	4	526,500	Blast Fce. Gas
A - 5	4	544,000	Blast Fce. Gas

Blowing Equipment

Number of Units	Kind	Capacity c.f.m.	Rated Pressure Lbs. Per Sq. In.
1	Turbo Blower	45,000	35
1	Turbo Blower	60,000	30
2	Turbo Blower	100,000 each	30
10	Steam Engine	33, 000 each	25

Reference 2

BLAST FURNACES (Contd.)

Auxiliary Equipment

High Pressure Steam:425 PSI from 3 boiler in No. 5 Boiler House with total capacity of 674,000 #/Hour.

Low Pressure Steam:Obtained from a central system for the north end of the Works, the system being supplied by the above 3 boilers (at reduced pressure) plus three other boiler houses (Nos. 1, 3, and 4).

Boiler Fuel: Blast furnace gas and pulverized coal.

Ladles: Iron - Mixer Type ladles - 2 @ 179 N.T. each
                                                 - 12 @ 190 N.T. each
                      Slab - Open Top Ladles   - 46 @ 18.6 N.T. each

Pig Casting Machines: 3 Double Strand

Sintering Plant: Four Dwight-Lloyd traveling grate sintering machines, each 3'-6" wide.

New Dravo-Lurgi Sintering Plant
First Operation - November 1960

Rated Capacity

Net tons per year	2,200,000
Net tons per operating day	6,500
(Potential of 8,500 net tons per operating day)	

Sinter Machine

Width	13'-2"
Number of Wind Boxes	183'-9"
Number of Wind Boxes	14
Grate Area - square feet	2,419
Machine Speed - feet per minute	5 to 20
Ignition Burner - B.T.V. per minute	150,000
Gas Mains - 12 feet 6 inches diameter	2
Bed Depth - maximum	16"

Reference 2

BLAST FURNACES (Contd.)

New Dravo-Lurgi Sintering Plant (Contd.)

Mixing Drums

Primary	- Number	1
	- Size	12 feet diameter x 30 feet long
Secondary	- Number	2
	- Size	12 feet diameter x 30 feet long

Main Sinter Pans

Number of Pans	2
Capacity - t.f.m. each - 35 inch vacuum	350,000
Motors - HP each	3,500
Fan Speed - RPM	720
Main Stack - Height	140'
- Diameter	18'

Sinter Cooler - Lurgi Circular

Rated Capacity - net tons per hour	444
Size - mean diameter	110'
Fans - 4-c.f.m. each	330,000

Building - Approximate Size

Main Building
Width	53'
Length	233'
Height	80'

Mixing Area
Width	53'
Length	77'
Height	120'

Reference 2

FLOW DIAGRAM OF NEW SINTERING PLANT

ALIQUIPPA WORKS

SWINGING STACKER

SCALE

STOCK PILE

SCALE

PRIMARY SINTER SCREEN

SINTER LOAD OUT BINS

SECONDARY SINTER SCREEN

STACK

STACK

STACK

HOT SINTER FEEDER & SCREEN

GAS MAIN

COOLED SINTER FEEDERS

COOLER DISCHARGE BIN

SINTER COOLER

PRIMARY MIXING DRUM

SINTER MACHINE

SINTER MIX SURGE BIN & ROLL FEEDER

IGNITION FURNACE

SINTER BREAKER

HOT SINTER FINES CHUTE

TRAVERSES

SECONDARY HEARTH HAULING DRUMS

HEARTH LAYER BIN

RECIPROCATING CAR

GAS MAIN

SHUTTLE

ROLL FEEDERS

ORE SCREENS

GROUND COKE

GROUND COKE

FLUE DUST

HOT & COLD SINTER FINES BINS

MISC MATERIAL & ORE UNLOADING TRACK HOPPERS

DOLOMITE

LIMESTONE

LIMESTONE OR ORE

BENSON CONCENTRATE

BESSEMER CLASSIFIER

BASIC CLASSIFIER

SCALES

COKE LOAD OUT BIN

ORE LOAD OUT BIN

FROM EXISTING FILTER HOUSE

GAS MAIN

RUBBLE SCREEN

BESSEMER CLASSIFIER

BASIC CLASSIFIER

TRACY

MECHANICAL DUST COLLECTOR

MAIN VENTURI STACK

MAIN FANS

RAW COKE STORAGE BINS

ROD MILL

LEGEND

MATERIAL

SYMBOL

C — COKE
D — DUST
FC — FILTER CAKE
HL — HEARTH LAYER
MM — MISC. MATERIAL — COKE, DOLOMITE, FLUE DUST, LIMESTONE, MILL SCALE
O — ORE
SF — SINTER FINES
SM — SINTER MIX
S — SINTER

Reference 2

Chapter 6

ഌ)രଃ

Blooming Mill

H e looked like a European fisherman. Cheerful countenance black
wool cap tipped off-center, heavy wool sweater, with thick rough
and callused hardworking hands.

My name is Eli Matish. Right now, I'm sixty-nine years old. I
came here in 1951, from the Ukraine. I was hired in the mill in 1952,
November 20th. My hourly wage was about a dollar twenty nine cents.
In 1986, when I retired, I was making eleven dollars an hour. I worked
for J&L for thirty-four years, the whole time but the last year. In 1985
I worked in the Tin Mill.

The Blooming Mill was where I spent most of my time, where we
made slabs from ingots. We took steel ingots from the soaking pit,
where it was red hot and rolled it into slabs, to billets, and to flats. So
the mill then took steel ingots made from iron and made that into smaller
and smaller units, and then it went to other places to be finished. I would
work on the hot bed where rolled slabs were placed, "hooker" to connect
the crane hoist to the load where the crane brought the slabs in and
"stamped" where the slabs were marked with a category and quality
control stamp.

First I worked in Labor Gang. Most of the time I shoveled coal or
coke from the old railroad cars to the soaking pit. The soaking pit, it

Eli Matish

was where ingots come cold and are heated up, almost white-red hot, and then to make the product we were supposed to make either blooms or slabs. There were days when I started we had about fifty people labor, twelve guys could shovel six railroad cars a day. We shovel about six cars of coal for the furnace a day. We used to open the chute there, just to clean 'um up, make room for more coke.

Well, stamping, first I went to hot bed to roll ingots. They would roll 'em most times, put small ingots through them. I think the largest was six-by-six, smallest is one inch and three quarters. I remember it was 180 inch long. Well, my job was to tell the clerk, and he would find out what grade of steel we wanted. Then he would give me the sheets to inspect and stamp the steel — high carbon, low, medium high or copper grade. And then I put a stamp on the steel, just like T, B or 2?

Well, in scarfing, we removed defects mostly from slabs and from billets, too. So we used a torch and just brazed off. Yeah, when I came through we started to use torches. Before me, we used to use a chipper hammer, but I never do that job. I just torch to take away a defect.

Oh, the day I retired was good because I was waiting for that time. Yeah, because I gonna move 'em from Aliquippa. I wanted to go to

Florida in 1980, but I didn't get my retirement, so I was waiting till that day come, but after 1986 it was to late to move anyway. Yeah, I took early retirement but at the same time I was put on disability 'cause I had trouble with my back and arthritis 'cause pain all over. Yes, that was from working in the mill for years my back hurt. I got hurt a couple times in the mill.

Well, when we turned the billets, we called them "ball-busters" the six-by-six, and we had to scarf them four ways, on each side. You had to turn them with wrench. I hurt my back. So you had to turn the billet. How much did they weigh? The billets weighed, something like a thousand pounds.

The worst day was when I worked on the Soaking Pit. They gave us boots up to my belly — pants and boots all together — and we would go in grease and especially when we put new bloomer in 1955 or '56. Something like that, because I started in 1952. About '55 they put new bloomer, it was bad work some of that work was hard in grease and hot in the summertime.

No, the whole time I go to mill I was tired, and come home. And I didn't feel good because it was hard work. I can't complain, because when I come to this country, don't know language, don't have school. I don't have tools, and to get me job in mill for me is something, a good thing.

I was 26, but I lived in Ukraine for six years. I was in Germany, England. I just come — no family — church sponsored me. Better life, everybody want to come to America. I don't care what people say or how much they complain, but the best country for poor man, for rich man, for young man, and for old man, this is the best country, the United States. Still right now, right now it's a little bit worse, but still is the best country, you know that. Most time I work because I get along with people, most of them I never had a problem with. At the beginning a couple of them young guys, they maybe joke about how I pronounce some words, but it was minor. But most time everybody respect me. No, I mostly want to forget, I don't want to remember, the bad, and good. I don't remember nothing good.

I have a wife and five children. There was a lot of bad shift work in the mill, 'cause you don't eat right, you don't sleep right, you were always tired. Family, well, they knew I had to work. I provided everything for them, send them to school, because of me I sent my kids to college, get them an education.

Blooming Mill labor force

My boy is working for government, and another work for Social Security, my daughter work for nursing, one is in Florida, one is a supervisor for handicapped children.

Everybody was sad on my last day, I remember that, especially the foreman. They play like a big shot, they thought J&L was going to keep them. I remember one guy, he was from Hookstown. He came up to me, it look like he was going to cry, it was sad. That was sad, the last days.

That was in June on a Tuesday in 1986. No, the mill closed before that, the Blooming Mill closed in 1985 on Tuesday in April. Then I was laid off for a couple months and was sent to Tin Mill. I was working in the electrical department. Vince Fuerher was the general foreman, and Nick Cavoulas was in the Strip Mill.

Well, the mill was hard work, but one of the things about the mill, like I tell before, you don't need education just a strong back. Like they use to say strong back and weak mind, you could make a living.

The end of the job

Blooming Mill rehabilitation

Scarfer cleaning slabs

Burning debris from rolls

Joe Biss

He appeared to be a studious man, bespectacled, gray flowing hair — could have been your high school history teacher.

I'm Joe Biss and live in Monaca, PA. I was born in West Aliquippa, and I'm eighty-three years old. I worked for forty-seven years, and was hired in 1930.

Well, 1930, The Depression was still on yet. When I worked in the Blooming Mill, you worked, they rolled the steel out. After they finished rolling the steel out, then you went home. They sent you home. At that time, everybody had it alike. There were three different turns, so if you went home early, then you came out the next turn. So that meant everybody was working, but not everybody was working all the time. Yeah, kept people employed, not like today. Today you're either in or you're laid off.

I think it was a good effect on people, I think it was a good way. That's the way we worked all the way, even when things picked up. Then, later on, they went slack again. Nobody got laid off. Things got so bad at one time. You didn't even have a regular starting time. You had to call the operator and you were like number eight turn, two turn, three turn, whatever turn you were and they told you what time you come out to work. You didn't have no starting time. That's when things really got bad. So nobody was laying off and everybody had to wait their turn, although you didn't get much work. Well, where we worked was always noisy and dirty in the Blooming Mill. I started out working in the Engine Room as a first oiler. The Engine Room, that was the steam engine to run the rolls. Well, had to oil everything up, all the rolls. That engine moved the rolls and the rolls moved the blooms and made them into billets.

Well from there in the Engine Room, they have what is called snow pumps. Big pumps, pump water through the mill and then out to the Bessemer. From there, they added what they called the pump man. He worked this pump. And then from the pump man, then you come back and worked down on the floor. From oiler you went to what they called engineer on the floor. Then you went up to the pulpit and you, the operator guy, worked the engine from up on the pulpit. Then you broke in on what they called an engineer.

So the engineer then sat in the pulpit and ran the steel back over the rolls with sparks flying. Yeah, he operated the engine and the engine operated the rolls. A large hot steel slab was reduced to a more narrow billet or a four-by-four bar and then we reduced it down to eight-by-eight billet or even they rolled slabs as finish products too. Plus, at that time, they had a bar and you rolled it to the Bar Mill.

After that they would take them to what they called the Hot Bed. Then they would take them wherever or the slabs would go out into the Slab Yard, or the billets would be reduced down in the Blooming Mill. Then go down to I-don't-know-how-many mills. I don't know how many stands they had to reduce that down. Then they would go to the Finishing Mills and make pipe or sheet.

Well, from working down in there I eventually worked up to the pulpit. I became what they call an engineer in the pulpit. Then after so many years they converted from steam to electricity, put those electric motors in, and tore those steam engines out, which made the job easier. All you had was one control. The other way you had two systems for forward and reverse. When they converted to electricity it was easier for me. The roll system was fifty to one hundred feet in length.

Then you would turn, go back and forth. They had a size, the ruler had a size to roll, like eight-by-eight or seamless. They rolled twelve-by-twelve, slabs, like thirty-five-by-thirty-five. Probably sometimes less than a minute to roll one billet.

Well, you were supposed to get to where like, you could easily get sixty an hour. Wait a minute, when you rolled seamless twelve-by-twelve — see, that goes down, sometimes you could roll seamless if they were hot — you could roll seventy an hour. Pretty fast.

You had an hourly rate plus a tonnage rate. Well, it was like certain sizes, you got a certain rate on certain sizes. In a day, I would say probably thirty dollar incentive, I'm guessing. You were probably making toward the end ten, eleven, twelve an hour or more, I would think.

Oh, a bad day was like, you would maybe only work four hours and go home. Run out of steel. They weren't making the steel, like down there. So we would roll what was in the pits and then you went home. You would go home and only get paid for four hours.

Shifts were tough on my family. We worked four shifts. Every forth week we worked a swing shift. A swing shift was three days daylight and two afternoons and one midnight.

There are four crews, twenty turns a week, so four fives are twenty, and they went down on daylight. We always went down on Sunday daylight. So there were four crews, each one got five days, so in order to work that way you had to work a swing turn week. I know, it sounds funny.

45 Inch Mill construction

Hot ingot from soaking pit

I retired in 1970. I retired while the mill was still working. It seemed like after 1977. I understand that's when things started to go down. I don't know, I think all over the United States it happened.

During the Depression, like I said . . . the president, when I first started in 1930, it was Roosevelt, and then they passed that National Recovery Act.. And then we had forty hours a week. That's when they had to put another crew on, since we worked forty hours a week. Before they only had three crews, and they worked like six days a week, sometimes five days. It all depended on how much work there was. That was before I worked up on the pulpit. You employed more workers and basically paid them a little bit less, but you kept more people working.

I think, I really think, that J&L was a good company to work for. I thought they were, they helped out. They did a lot of things for Aliquippa, like putting in a football field and stuff like that. Well, I think everybody, I thought, was treated fairly. I don't think anybody was actually hurt. They treated them fairly. It seemed like where I worked they were all pretty good guys. There was daylight and two afternoon and one midnight shift in a rotation over and over for forty years.

Old 44 Inch bloomer turning ingots to slabs

Hot ingot on blooming machine

Bar and Billet Mill

Blooming Mill post demolition

ALQUIPPA WORKS
BLOOMING AND BILLET MILLS

Production Data

Soaking Pits

    Six 3-Hole)
    Four 4-Hole)   4,937 Sq. Ft. Heating Area
    One 5-Hole)   2,400,000 Net Tons Heating Capacity/Year

Blooming Mill

    A-1 44" 2 High - 1 Stand   1,989,000 Net Tons/Year

Billet & Sheet Bar Mills

    1-15"   (21" 2 High - 6 Stand)   Continuous
           (18: 2 High - 6 Stand)   900,000 Net Tons/Year

    1-Sheet Bar & Skelp 18"   (21" 2 High - 4 Stand   )
                        (18" 2 High - 8 Stand) Continuous
                        (2 - Stand Vertical Rolls)

                        425,000 Net Tons/Year

Operating Statistics

Mill	Product	Tons Per Hour Rated	Actual	MMBTU Per Ton
A-1 44" Blooming Mill	Slabs & Blooms	235	262	-
21" & 18" billet Mill	Billets	109	-	-
21" & 18" Sheet Bar Mill	Billets & Skelp	51	-	-
Soaking Pits	-	-	-	.709

Equipment Data

Soaking Pits

	No.	Type	Main Fuel
Single Pits	5	Recuperative	By-Prod. & Bl. Fce. Gas
Single Pits	18	Recuperative	By-Prod. & Nat. Gas
Single Pits	14	Regenerative	By-Prod. Gas
Double Pit	1	Regenerative	By-Prod. Gas

Reference 2

BLOOMING AND BILLET MILLS (Contd.)

Mill	Make	Description	Size of Rolls (Dia. & Length)	Drive	Speeds BPM Base	Top	No. & Size of Motors
A-1 44" Bloom	Mesta	2 Hi-1 Stand	43"x 90½"	Elec.	70	140	4-3000 HP
Billet Mill	Morgan	(21" 2 Hi-6 Stand)		Var.	Elec.	-	-	1-5750 HP
		(18" 2 Hi-6 Stand)			Elec.	-	123.3	1-5750 HP
Sheet Bar &		(21" 2 Hi-4 Stand			Elec.	-		1-5750 HP
Skelp Mill	Morgan	(18" 2 Hi-8 Stand		Var.	Elec.	-	123.3	1-5750 HP
		(2 St. Vert. Rolls						

Auxiliary Equipment

Billet Mill

1 United up-cut shear
1 Morgan crop shear
1 Morgan flying shear

Bar Mill

1 Morgan flying shear
1 United up-cut shear

Hot Scarfing Machines

Make of Scarfers — Linde
Fuel — Natural Gas
Maximum Speeds - FFM — 180

Product Range	Bloom Scarfer	Slab Scarfer
Width - Inches	4-14	6-53
Thickness - Inches	2-14	2½-14½

Chapter 7

℘ℭ

Strip Mill

My father was a foreman, serious man, all business, as he discussed his life and times. He is now 95 years old, and still loves to sing Polish songs, both religious and folk.

My name is Marion Prajsner and I live in Raccoon Township — Beaver County, PA. I am now 69 years old and was born in Poland, not too far from Krackow. I emigrated to the USA in 1935 at the age of seven, at the height of the Depression. I attended a parochial school at St. Titus parish in Aliquippa, and in 1943 we moved to Hopewell Township where I finished high school.

I started working part-time on weekends and during summer vacations at J&L Steel in Aliquippa from 1946 to 1948. I worked as a laborer in the Seamless Tube Mill. In 1948, I transferred to the Tin Mill and was hired to pack tin off the shears. I worked this job for one year and then took a job as a material man, which was ordering and getting covers made for packing the sheared tin.

In 1950, I again changed jobs and started to work as a Roll Grinder in the Tin Mill Roll Shop. In the Rollshop, we ground and reconditioned rolls for the Tandem Mill and the Temper Mills. This included grinding work rolls and back-up rolls.

While I was working in the Tin Mill I put in a suggestion on a new type of finish for the bridle rolls for the electrolytic tin line and pro-

Marion Prajsner

ceeded to adapt this finish to the surface of these rolls. It looked like a finish on corduroy trousers and so I named it a corduroy finish.

Before I designed this finish, the bridle rolls on the tinning line were changed every four days because of slippage and stripwork, which damaged the strip edges against the tanks, thus causing the damaged strip to be sheared down to the next width down. This caused a great amount of damage and cost. After I grooved the rolls to the new finish, they were installed and lasted eight weeks and then were taken out and the top roll was installed in the bottom and bottom roll was put on top and the used for another four weeks, thus saving a lot of down time and production.

At the particular time, the award for a suggestion was calculated at 10% of the annual saving to the company, or $10,000 as a top award. Mind you that there was a tremendous amount of savings in down time — production — and prevention of damage to the strip. Guess what the award was? It was $10.00 and only $8.12 after taxes.

When the general foreman asked me how much the check was for and I told him, he blew his stack and told me to take it back to the superintendent and throw it down on his desk and tell him to buy enough of rope to hang himself. He couldn't understand why I didn't get the top

award with all the saving that resulted from this suggestion.

I only wanted to give you an example of how the working man was treated and let you know that supervision took credit for a lot of the suggestions the workers submitted.

In 1958, I transferred to the 44" Hot Strip Mill. I was asked to transfer to the new mill because of my experience and only under the condition that I would be the sec-

Marion Prajsner in 1950 (22 years); mother, Anielo, (born 1910); father, Joseph, (born 1903)

ond oldest in seniority, since I was being laid off periodically, I thought this would assure me steady work.

I did not work on the mill in the Strip Mill but in the Roll Shop.

We ground rolls for the Reversing Reducing Mill. The six stand Reducing Hot Mill, and we ground edging rolls for the Strip Mill and 14' Structural Mill.

I also ground accumulator shafts that were 22.5" in diameter and sixteen feet long.

There were also times I ground and polished the journals of the back-up rolls to a mirror finish so that the babbit bearings would last a very long time.

The various roll shops throughout the mill were under a separate superintendent and the different mills would pay for our service, although the shops themselves were within those various mills.

My first superintendent was Mr. Hoffman and the assistant super was Mr. Currey. These were men that were very good to work for and were very understanding.

The second pair of superintendents were Mr. Winkler and Mr. Baker. Mr. Winkler was a very intelligent person but was all business and Mr. Baker was a regular guy that came up through the ranks and treated me well and we kidded each other a lot.

The worst day I had down at the mill was when my buddy had a heart attack and died in my arms. This was in the Hot Strip Mill Roll Shop office and it was on the midnight shift.

The healthcare was good in the mill and at the health clinic and if further treatment was needed they sent you to South Side Hospital.

I was quite satisfied with my work and was treated well and with respect. I always tried to do my best and put out a product that was done accurately and with pride in my work.

The one main regret that I have is that I did not go to college. I had the opportunity in 1958 to go to the University of Pittsburgh for engineering. At that time, Jim Forbes was the head of the Training Department and he was after me for a whole year so that I would go to school. He said the company would pay my full tuition and the only stipulation was that I would work for the company for five years. I just started building my new home and my wife was very ill and there was no way that I could attend school. Oh well, such is life.

J&L was a very good company, but when LTV took over they cut maintenance to the bare minimum. They did not service the machines and only repaired them when they broke down. On my machine I made sure that I serviced it myself because if the motors were not oiled or the filters not cleaned my machine would not function properly and the one thing I didn't need was a machine that didn't run well.

Many times the electricians, the pump men and millwrights weren't to happy about our oiling and maintaining our machines, but if you wanted to keep up with the mills and supply them rolls, you did what you had to do to keep them rolling steel because that was our bread and butter.

I recall a very nice person by the name of Mike Begg. He was my foreman and you couldn't ask for a better person that treated everyone very well and with respect and he would stick up for his men. I miss him very much. In fact, I miss the whole gang.

Where I worked in the Roll Shop, at times there was a lot of smoke and I had a headache almost everyday from the dust off the mill and the draft. In the winter, especially when they opened the main door to shift railroad cars into pick up the scrap, all that wind would blow in and sweep in where my machine was and the dust and smoke was so thick that at times, I had to shut down my machine because I couldn't see or breathe properly.

In the winter, I had to wear earmuffs and a heavy jacket and my hands were always cold because of the water spray from the coolant and the steel frame of the micrometer.

The superintendent would come to me at times and seek my advice when ordering grinding wheels, as he was confident in my ability to select the right hardness and density. The hardness and density was very important because if they were not right the wheel would develop chatter or if too soft the well would breakdown and not carry a full pass and wear down and cause an uneven surface.

When a grinding wheel salesman would come, the superintendent would come to me and say, Marion, you take care of it, you know what is needed, please take care of it.

As a consequence of this, some of my buddies would be a bit upset because I was asked to take care of the salesperson. It wasn't my choice but I felt that if I selected the right grade of grinding wheel, it would be better to work with.

I used to like the lunch breaks. They were a time of relaxation. We used the office to eat lunch and it was a full twenty minutes of jokes and laughter. Those were happy times. You would come out of the office holding your stomach from laughing so hard. Some of our fellows were really good at telling jokes and it sort of lifted your spirits and gave you more incentive to go back to work.

The workplace was dangerous at times. The floors were oily and greasy. We hooked up very large rolls with 1 1/4" cables and some times the wires in the cables were frayed and you would get puncture wounds in your hands, thus causing infections, or rolls overhead and some times cables snapping or the craneman swinging the load. There were many things you had to look for to be safe.

I have to mention that I had good cranemen who were very careful and we got along very well and we worked as a team. Some of my fellow workers used to wonder how the craneman used to be above my machine exactly when I needed him and not necessarily over their machine when they needed him, and I used to tell them that if you treat him right he will treat you the same way. I always knew which side of my bread was buttered.

My father, Joseph, worked in the Rod Mill. The work wasn't too steady, a lot of down time. The Rod Mill, through the years never really had a lot of orders except during the war. I had a little sister that died at three years of age in 1939 and in 1948 and 1949 my mother and father had two more daughters, Teresa and Jane, and I love them all. Dad was an accomplished violinist and I used to love to listen as he played the different polkas, obereks, waltzes, conchertos, sonatas. He also loved compositions by Fritz Chrisler. I would watch him in front of the mirror perfecting his stance and positioning and bow work. My father made his own violin in Poland because they could not afford to buy one, and I'm proud to tell you that it was an instrument of high quality.

In 1987, my father and I returned to Poland to visit our family and naturally his nieces and nephews asked him to play. The whole Prajsner family in Poland is musically inclined. One of my cousins played the violin quite well and was interested in how dad played. So with a little urging, at age 84 he stroked the strings on the violin and what came out was the sound as though it was a Stradivarius — beautiful — soulful and full of feeling. My cousins were brought to tears and remarking, uncle, you are a true Prajsner, a true musician. We stayed for six weeks and every evening we all gathered in Polish fashion for music and singing.

The last day of work at the mill was spent under the Soaking Pitts in the Blooming Mill shoveling scale. When the Strip Mill shut down I was laid off for six months and then called back and put into the General Labor Gang. We cleaned scale out of valve pits, used jack hammers in the Tin Mill and hammer drills. I came home one day and went to sleep and awoke and heard my wife laughing and I asked why and who put on the light and said "Marion look at your hands" and I did and what do you know, they were shaking as though I was still using the hammer drill. To make the story short, on the last day, one of the fellow workers came to me and said that he heard that the first wave of Strip Mill workers with enough seniority were to be given the chance to retire. After hearing the news none of my buddies were too eager to work, but we still did a bit more. When we came up for lunch the foreman informed us that this is the day for those of us with enough seniority, will be offered pensions. Believe me, after that announcement there was very little work done because we were just too excited, and as far as the company, they were going to shut the place down anyhow so we just didn't have any guilt about the last four hours. I had given them thirty four years and nine months and it was time to move on.

We lived on Kiel street in Aliquippa and my father used to go down to the "Y" and mingle with the workers and union people and I saw a bit of what was going on. The struggle of the workers to organize and the town police would chase and arrest them. As you probably know, the whole town and police at that time were controlled by the company and to defy the company would mean you would lose your job.

I have been a union man since I started work and will probably die supporting the union and the workers and retirees.

You know, as far as I'm concerned the mills in Aliquippa would still exist if we would have had the cooperation of the government, not to let the company just shut down plants indiscriminately and raid their pension plans. When LTV bought Youngstown Sheet and Tube and Republic Steel, it didn't make any sense because they had so many Bar Mills, Seamless Tube Mills, Strip Mills that they had to shut a good many down.

It had been alleged that the reason they bought Youngstown Sheet and Tube and Republic Steel was to raid the pension plans and buy into the Aerospace Industry, but I say just piss poor management and stupidity. To further press my point on management stupidity is to tell you about something that happened after LTV bought Republic Steel.

There is a plant in Indiana, a foreman Republic Steel Plant where they couldn't get the proper production on their Blooming Mill, so they sent a crew from the Aliquippa Works to help them get more production and wouldn't you know that in six months they helped to double production.

Now you tell me, is it stupidity or what to shut down a plant where such superior workers are so highly productive and to keep one open with less skill?

Aliquippa Works of LTV used to make some of the best steel, was noted world wide for their drill pipe, our Blue Ribbon Steel Pipe.

I have nothing but sadness, not for me but for the young people, there are very few good paying jobs in our area.

I can not see how the government could let us down by not stopping these indiscriminate shut downs that are killing whole towns and cities. Why the Congress does not stop these people who take over companies and break them up, sell and then take tax write offs, this is beyond me. Every time I read or hear of another merger or acquisition, it makes my blood boil because it probably means some more loss of jobs.

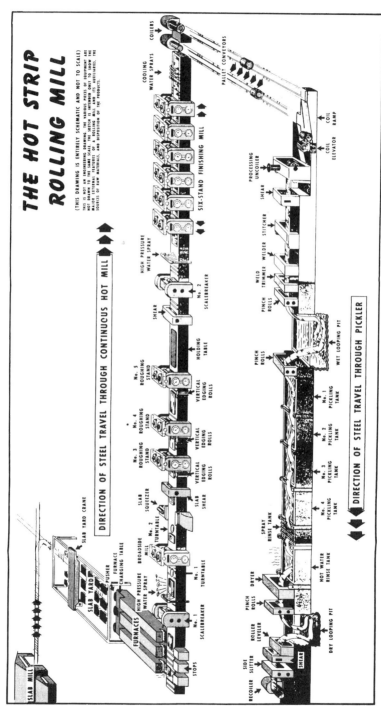

Hot Strip Rolling Mill Schematic   (Reference 3)

Columbus Day parade on Franklin Avenue

Hot Strip Rolling Mill

Rolling Mill repair

Rolling Mill repair

44 Inch Strip Mill control pulpit

Strip Mill demolition

ALIQUIPPA WORKS 44" HOT STRIP MILL

Production Data

Annual Capacity   1,200,000 N.T.

Equipment Data

Mill	Size of Rolls (DIA. x Length) Work Roll	Back-up Roll	Size of Description	No. And Speed Motors-HP	Strip BPM
—Scale—	12"x14" Face		Bottom Drive - Overhung Roll	2-500 HP	-
Reversing Roughing	45½"x44" Face		Universal 2-Hi	2-3000 HP	115/1100
Attached Edger	26"x17" Face		Worm Driven	2-375 HP	-
Crop Shear				1-300 HP	-
Finishing Scale Breaker	16"x44"		1-75 HP	240/480	
#1 Finishing	22½"x45"	46"x42"	4-Hi -1 Stand	1-3000 HP	209/400
#2 Finishing	22½"x45"	46"x42"	4-Hi -1 Stand	1-3000 HP	373/400
#3 Finishing	22½"x45"	46"x42"	4-Hi -1 Stand	1-3000 HP	530/1060
#4 Finishing	22½"x45"	46"x42"	4-Hi -1 Stand	1-3000 HP	735/1471
#5 Finishing	22½"x45"	46"x42"	4-Hi -1 Stand	1-3000 HP	881/1762
#6 Finishing	22½"x45"	46"x42"	4-Hi -1 Stand	1-3000 HP	1178/2356
1 Edger			Integral	2-150 HP	136/270
2 Edgers			Integral		587/1175
2 Downcoilers	24½" I.D.		Expanded Mandrel	2-17 HP Each	Pinch Rolls
				2-75 HP Each	Mandrel
				1-200 HP Each	Wrapper Roll
				4-15 HP Each	

2 Slab Heating Furnances   150 NT/Hr. Each

Reference 2

Section III

South Mill

Chapter 8

ಶಿ಄ಞ

14 Inch Mill

H e certainly had the best nickname — Hillbilly Jim — he said, with pride. His youthful appearance, and enthusiasm was in contrast to bilateral hearing aids, a remnant of hearing loss due to the continual steel clatter of flat and angled iron running through steel roller assemblies.

My name is James R. Coe. Hillbilly Jim, they used to call me. I'm sixty-one years old born in Alexander, West Virginia.

I had thirty-six years of service. I started September 19, 1953 and retired September 1, 1989. Not too much, about three thousand six hundred dollars a year was what I made when I started. I wasn't getting full-time, was getting twenty-five or twenty-six thousand dollars. Well, when I started, I started in the Seamless Tube. I had thirteen and a half years in there, from 1953 to 1967, and on April 3rd I went to the 14 Inch Mill.

The 14 Inch Mill made beams and flats from a billet. Some billets are square, some are flat, all different sizes. I don't know, ten or twelve feet long I imagine, not too much longer than that. Could be a little longer. You reheat those, you run them through the furnace. They come from the rough mills and they had finishing mills, number twelve was the last mill, that was finished product that came out of number twelve.

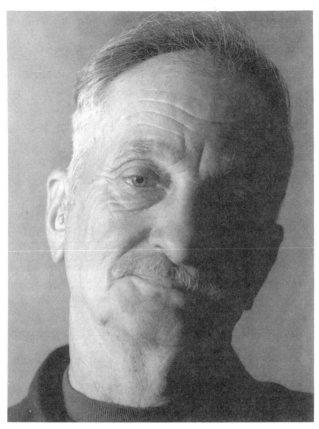

Jim "Hillbilly" Coe

So you had to heat the billets and then you ran it through a rough mill to get it to the right size. Then the finishing mill made the size and shape. You make beam, rim, angle, flat whatever you goin' to make, you know. Those products were used for construction. They build stuff, see, the beam, a lot of them went for the frames for the trailers. Beck bought a lot of beams.

My last job, I moved up. I was a hot bed operator. I had that job for about nine or ten years. I transferred the steel from the hot bed, I fed to straighteners and from the straighteners. They went on the line to the shears, and the shears cut the steel and they shipped by truck or railroad.

I was an hourly employee. Some jobs it was a hundred percent bonus, some were fifty percent bonus. You had roll so much to make. More you rolled, more money you make. Sometimes you could make seventy-five or eighty dollars, in addition to your hourly wage. Yeah, sometimes you could make more incentive than you made hourly. I had a eight point job. I only made ten something an hour. So you could actually double your salary with your incentive.

Yeah, we, like, work a holiday and have a good day, get double time and a quarter, you got double time and a quarter on your incentive, too.

You could make some money on a holiday. See I was on turns we use to break a lot of records, you know. The more you roll, more you made. Always break the old records you know, roll more. Tonnage rolled, yeah, that was tonnage. It was a good product, see, the scrap you didn't get paid for. If you rolled scrap iron all day, you didn't make no money, you only made your hourly. You only got paid for good product. They had an inspector down there.

The best day we didn't have no trouble gettin' real good rolled steel. See, some days, a lot of times harder you work, the less you make, if you had troubles, you know. Good days you had less trouble with the more production you could get. I guess like anything you could have good days and bad days. Lot of days you fixed up what was breaking down so people weren't breaking down. You'll wear out. You buy a new car, it's goin' to break on you.

Worst day, most of the time when you really had a bad day. I seen there would be two or three days before they could get things right sometimes. Sometimes for a long time we wouldn't have no problem having mill change come right on with a new size and then some days it would. It seemed like rim was the worst. That was for car wheels.

We rolled for Firestone, they rolled a little bit of everything one time, they made a cutter bar. They only roll that about once a year, that was for cutter bar on the mowing machine that you cut hay with, you know. It was construction, they would buy anything you need. At least you got enough to keep the wolf away, you made a living.

Yeah, the noise, it was loud. That's where I ended up being. I think Seamless was the worst place, that pipe banging in steel cradles — suppose to be wood, but the wood wouldn't stay in, it wear out. 14 Inch Mill was noisy, but I think the Seamless was the noisiest. I got two hearing aids now. I still ain't got my right hearing. Yeah, the mill was a good place to get killed. You would think they would have more accidents than what they did. They never got too many people killed down there.

Most men were pretty safe. You had to be careful, you know, but it was a danger. You see how that mill operated, the 14 Inch Mill, when they roll a special angle when the cobbles come up they could come right up and cover somebody up. People be standing on those mills and a bar would come right out of there, you had to jump and run. My family . . . you learn to live with that, that all goes with life.

I don't know. I would rather have had steady shift myself instead of swing shift. If I was going to have steady shift I would of took the four to midnight. That is the hottest in the summer but the warmest in the winter. Yeah, liked the four to midnight shift the best, eat the best, feeled the best. I would take four to midnight over daylight. I hated daylight. I seemed like I just couldn't handle it, didn't feel good till time to go home. I hated daylight.

Last day of work, I guess we done just as good on the last day as we did on the first. We made money. While we were there, we made money for ourselves and made money for the company. Good to work for the company. I had no complaints. I must've liked it, had thirty-six years in there.

Well, mills are all right to make a living, but I say it ain't no good for your health. Yeah, dirt, dust, a lot of dust, you know, you bound to breathe that steel dust. See, where I made my mistake, but see, when I went to work in the mill, I should have had cotton in my ears but I didn't. They didn't give you're protection, they waited until you got to you couldn't hear. They made all wear earplugs, you know, after you couldn't hear no more. Should've had that the day I was hired then. But if I would have been smart I would've put cotton in my ears that would of been just the same, but I didn't, and I lost my hearing.

It was a good place to work. I have nothing against it, it was a good bunch of guys. I was eighteen. I was eighteen when I came up from West Virginia. Yeah, that's why I come here, I come here looking for work. I had an offer at the time to move to Ambridge. I come up, my aunt told me at first two weeks I be on vacation and after that I pay rent. Rent and board and I stayed with her till I got a pay day or two, and I moved out on my own, and I've been on my own ever since.

There's a lot of difference between Aliquippa and West Virginia. It was people I never knew was around. I came to Aliquippa and, see, I was a country boy, I come off the farm, and hadn't seen to much till I came to Aliquippa. There I seen a lot. Well, there had never been too many times and I still don't like city life, I still lives out in country like, I would never live in the city.

Good, Aliquippa was boomin' one time, like around Christmas time. Get off work at four, it might be seven before you got up Franklin Avenue through town. That's how busy town was, when the place was really working you know. There was a lot of people down there when they use to work their twenty turn schedule all the mill.

Yeah, twenty turns that meant work to every shift but Sunday daylight and if they worked that, too, that was extra days. You got six days that week if you worked Sunday daylight. When I was a single man I use to get Sunday daylight, and it seemed like everyday I wanted to work late, I could never get up on time. It got so they say, you be out about eleven. I say, yeah, I be there. I never had no problems with the foremen.

I liked the foremen, the management was good. Got along with them all. They had a job to do, I had a job to do, that's the way I looked at it.

14 Inch Mill construction site

Junior I beam production

Sorting and stocking steel flats

The "tunnel" that led to the main gate

Main parking lot

Dennis Frioni

He seemed younger than the rest, quietly and objectively describing the events that resulted in the loss of his job and livelihood before his prime.

My name is Dennis Frioni and I'm from Baden, PA. I'm forty-five years old and was born in Sewickley.

I worked for twenty-one years, before it was sold, and was hired in 1969. I was making two dollars and thirty-five cents an hour. August 1987 was when the mill went down. It didn't shut-down, they just idled it and the sale was completed November 6, 1987. Then, in November 1987 the mill was sold to three partners, Seth Breedlove, Jim Howell, and Carl Snyder. Two former managers and an outside investor.

I am still working, and make fifteen dollars an hour. This is at J&L Structural Steel, the old LTV 14 Inch Mill.

The 14 Inch Mill itself makes structural beams, shapes and plates. They take a billet which is raw steel, reheat it and reform it. Previously it was made in the Aliquippa Works. It was either a Blooming Mill billet or a cast billet. Right now, we have about four suppliers of billets come from Bayou, Roanoke, or Florida Steel. They're trucked in or shipped up by barge. The smallest billet is a four-by-four by twenty foot, and the largest is six-by-seven by thirty foot.

It goes through a reheat furnace and then it enters the mill and it comes out the mill a finished product. The mill is a series of rolls. Different combination of rolls that reduce steel to the shapes and forms. The Junior I-Beam was invented by J&L Steel, they had the patent on it. It's a light weight beam used mostly for manufacturing housing. The 14 Inch Mill was started back in 1929. They made different sizes and shapes of flats, they used to make a truss, they made bullet casings during the war, they made cutter bar, they made precision plates for cold rolling for the automobile industry.

I couldn't tell you, I'm not too familiar with it, but either in the early 60's or maybe the late 50's, J&L developed the Junior Beam which was a light weight structural product. At that time they were working with light weight structurals — angles, flats, plates — and the late 60's or early 70's there was a boom for the Junior Beam. They most exclusively went into the Junior Beam business, giving up the other business, keeping a little bit there. The mid 70's when I went down there to the 14 Inch Mill, the market bottomed out, people were building stick houses, trailer industry was going down, and it was from '74 to really '87 the Junior Beam market was pretty well low, at its low point.

When we transferred ownership and went with the new company, we started pretty decent and the market picked up and it's been. We went through, this is our second cycle, from '87 to about '90 the market was good. '90, '91 the market was down. The mill was down to about forty hours a week, sometimes less. '92, the market shot up and we've been rolling, the mill has been rolling six, seven days a week since '92 to present. Now we are back in our slow period, we're starting to slow down now.

Structural steel, at one time it was used for residential, but mostly it is used for manufactured housing, trailers, modular homes. The channel we use for stair steps, stair guides. We slit the beam which makes into a T. They use that in the marine areas and ships and submarines and that. Presently, we are not rolling flats. We got out of that business because the tolerances are a little bit too tight right now, and the market for structurals are up. We roll guardrail.

Yes, the tolerances are too tight, the specifications that are required by the company. We were having problems meeting the specs with them, and we were getting a lot of rejects. We had the same problems with the older mill. They start investing the money into it. They bought new guides and then the market got kind of soft. We roll the guard rails to six inch wide Y-flanges used for guardrails, and the 3-inch that is used for guiderails.

I'm an electrician in the 14 Inch Mill. I maintain all the equipment down there. A typical day on a rolling turn is that you are building up parts and, therefore, breakdowns. If there is a breakdown, piece of machinery goes out, you fix it, get back on line. You are basically on-call on the off turns, and you are for predictive and preventive maintenance.

Yes, when I got out of the service in '72, they had an apprenticeship at J&L. I applied for it and received it. I received two years training then, in 1974, and was sent to the 14-inch Mill and have been there since. Two years full-time was 1,040 hours, step grade up to four grades, and then you tested and you received your pay rate.

My best day was when we were sold. That was my best day, when we were sold. I was there, I was part of the negotiating team with the union. Put the contract together. What we did was after it was negotiated and the sale went through, we just went from day one and ended your LTV career to J&L Structural Steel.

There was a buyer before that, a guy by the name of Yung Paik. We negotiated a contract with him, the union and Yung Paik, and Carl Snyder and Jim Howell was part of that first thing. The contract was negotiated, signed, ratified, and one day there was — in fact, the governor came down and even presented — Governor Casey presented him with a check for four million some dollars to help him get started. About a week later, one of the managers come down, had everyone come out of the mill and said that the guy backed down. No reasons, just backed down.

Rotating turns are rough when you're raising a family. Well, when your kids are in school and you're on the afternoon turn, you don't see your kids for a week at a time. You are always constantly running. You basically . . . you have to find time to make quality time for you and your family.

Well, we work on anything from a DC motor, overhead electric cranes, motor generator sets, rectifiers. It's usually a breakdown — they could break a shaft, they could lose a motor, you go up and change an armature or repair shaft. We have eleven electricians now. Normally three men on afternoon and night turn, and the rest are day turn. Actually, it was smaller before the mill shut down. It was smaller, because at that time we had eight electricians when we were LTV, but we also had a general maintenance department which comprised of RIW's, who were repairman iron workers, boilermakers, pipe fitters, welders, carpenters, electricians. We do everything and service all of those.

No general maintenance, now, we have our own maintenance. Whatever we can't perform is contracted out, they bring in outside people. Working for the new company, the morale is up and down. There were a lot of changes. We were just sold again, and the trust between the people and the new owners and the corporation is no longer there. Before we were a locally owned company — three owners, hands-in, hands-on. Now we are . . . the owners are still there managing, but they're controlled by a holding company.

It's back to like working for LTV versus J&L Structural. Well, they're going to have to have more hands-on ownership and get down and meet the people and get back to where we were.

Bar and Billet repair

Try working there for a summer, anyone who is going to school or thinking about not going to school. Apply for it and try and see what it's like. They would go back to school.

I don't think people understand the environment that a steelworker works in. All you hear is that a steelworker is overpaid. A lot of the money is due to the conditions, the hazardous conditions and the environment that you work in. So whether early or late, the mill affects your health. Yes, were hot, were cold, never comfortable in there, constantly in danger.

Angle iron and I beam

Rolled flat steel for tire rims

Former steel yard

ALQUIPPA WORKS
14" BAR MILL

Production Data

Annual Capacity   345,000 N.T.

Operating Statistics

Product	Size Range
Junior Beams	6", 8", 10" and 12"
Joists	10" and 12"
Light Beams	14"
Special Channels	10" and 12"
Standard Channels	3", 4", 5", 6" and 8"
Angles	1/8" to 1" thick 1-½" x 1-½" to 6" x 6"
Flats	1-½" x 3/16" to 15" x 2-9/16"
Rim Sections	5", 6" and 6-½"

Equipment Data

Make	Description	No. & HP of Motors
United	2-Hi - 12 Stand Continous	
	3 Roughing Stands - 18"	1-3000 HP
		(1-750 HP
	3 Vertical Stands	(2-200 HP
		(2-1700 HP
	5 Intermediate Stands - 16"	(2-2000 HP
	1 Finishing Stand - 16" or 30" Univ.	(2-2100 HP
	1 J&L Furnace	
	300' Hot Runout Table	
	256' Mechanical Hot Bed	
	2 United Shears	
	2 Ryerson Cold Saws	

Reference 2

The Mill

Aliquippa Works
Bar Mill

14"

BILLET HEATING FURNACE

BILLET

16" TWO HIGH CONTINUOUS STANDS

VERTICAL EDGING STANDS

14" TWO HIGH STANDS

UNIVERSAL STAND FOR BEAMS

COOLING BED

Reference 2

Chapter 9

ℰ᠙ℭ℞

Welded Tube

H e was a stocky, solid man, with thick speech nearly impossible to understand, red, worn, flannel shirt with huge forearms and dense callused hands.

My name is Elmer Cumberledge and I'm sixty-six years old. I was born in Monaca, as far as I know, am, like, part German and part American, or something like that. I worked for thirty-six years, and was hired on May the 8th in 1947. In 1947, I was making eighty cents an hour at the time. I retired in 1983 and ended up making, it was like, a little over ten dollars an hour. Welded Tube was the department I worked in at J&L. The Welded Tube department mainly made steel tubing. At first we took like flats and folded them over, added a weld, and then the system turns over to the continuous weld, which took rolls and produced pipe without folding.

Over the years — you don't want to know about that, you know — but I didn't think I would ever last that long. So the Welded Tube took through a round and then it came together to be folded, as it was welded, or it was more or less like a seam on the order of an electric weld as it came through the furnace. The Welded Tube department sold galvanized pipe to the gas companies and, like, all kinds of hardware stores for couplings, nipples, and things like that.

Elmer Cumberledge

It was laborer, my first job in the Welded Tube. We did a little bit of everything on the labor gang. We'd sweep and clean-up and stuff like that. Well, I moved up and ended up on the finishing bench, I ended up there. I ran a coupling and screw-on machine on the finishing bench. That came over like that, and picked up a coupling and put on the pipe.

One thing I remember, one day when I fell in the mill and came down about fifteen feet. When I came down eight five-inch pipe rolled off of that stack, all the way down to the ground floor. See this scar right here. Luckily that was all I got out of it. That was really something the way that happened, you know what I mean. So the Lord was with me.

The other hardest thing, I'll tell you, the whole time I was in there I was on my feet continuous. I more or less didn't take it easy any time, to tell you the truth. So you would work eight hours, twelve hours, sixteen hours on your feet. More or less on your feet. I don't know how that was done. I couldn't tell you. You know what I mean? I ended up there thirty-six years. You'll probably won't believe this, but I only missed two days in thirty-six years? Actually on my own in thirty-six years. No, I would, I mean would, actually taken it off on my own personal days. Right, now I ain't talking about sick or what not, but I was off for six weeks whenever I was injured.

Well, there's quite a few friends. Bob Buday — I don't know if you know him or not — he passed away. Well, the way he used to work hard, used to do in the mill, oh yeah, he was a good guy to work with and helped me a lot. Well, I'll tell you, my wife didn't actually like it when I was out till midnight, but what are you goin' to do?

Oh yeah, I got four kids. Four children. Well, they didn't say nothing about it, I guess. I started there at midnight, I ended up on midnight when I retired.

The midnight shift at the mill was different than daylight actually because lots of time you couldn't get sleep. No, I actually could never sleep on midnight. Some guys could sleep in the mill. I'll tell you, a lot of guys did and that's true. Many times I stood by that machine, more or less they would just stand there half asleep. It was a wonder that I didn't get hurt. I always made it.

Well, I think if I could change a few things about the mill, I probably would but, like I say, it's hard to tell.

I'll tell you the last day, that day, I could've killed the boss that day, because we were on buddy relief. Your buddy has to come in before you go home. That day, well, on my job, anyway, I had a buddy who ran crane. Well, I was riding with him, so on this last day, Walter Waldo, I think his name was — no wait a minute, this was a new boss, I forget what his name was now — anyway, he came over. At first he said, oh yeah, you can go on buddy relief, then he come back out — you can't go until your buddy comes. Okay, that was all well and good. At three fifteen he comes and tells me, well, you can go if you want to, and I really think you will. Like I said I had a ride home, but he had already left then because he was ready to leave. We know'd that, and actually he had got me peeved off that day. So then I missed my ride. He said go home and don't get paid, so now I missed my ride? Yeah, that was my last day.

Well, put it this way, you couldn't tell the younger guy. Younger guys *know more than you do*, believe me. At least, they figured they knew more than you know. Anything, they know more than what you know. One time I was working on a job down there, for thirty-five years, thirty-six years, and the boss came down and he said I want this done this way. So what I did, I nicely shut the machine down and I walked away, and I said show me how you want it done after I've been on it for thirty-five years. He just walked away, he didn't say nothing. Right. He just walked away.

Bar storage yard

Young John Turkovich (1940)

John Turkovich

He had an almost regal appearance; eclectic eastern European dialect at times. Surprisingly, his appearance had changed little over the years since he stood as a youth in front of a 1940 coupe.

My name is John Turkovich and I'm seventy-two years old. I was born in Aliquippa. My ancestors are from Yugoslavia, I'm Serbian. Well, counting my service, I worked for forty-two years.

I worked in the Welded Tube Department where I was hired in 1941. My wage was five dollars a day, when I was hired. March 1983 was when I retired. Well, I was on the piece work. It was variable, around twelve an hour.

The Welded Tube Department made different tubing, well, I worked on the finishing part. I didn't work on the furnace. Well, the furnace, they made the pipe. Well, when I first started there, they use to make pipe out of slabs. Then years later they converted it into the coil method. I worked in the galvanizing department but I threaded pipe. Well, I threaded from half inch up to two inch on one machine, and then later on they got these new, automatic threaders. It was all automatic. It threaded it from half inch to four inch pipe.

Well, a lot of it was used for gas lines, a lot of it was regular standard pipe like the waterlines. Then special pipe made the gas lines, and then they made this conduit for electrical wires. Well, labor was my first job title when I first went in the north mill, the blast furnace. That was going in and out of the blast furnace when they overhauled the blast furnace, replacing the bricks. We used to work on shifts, one gang would go in there for about ten minutes and we wore shoes, wooden soles about six inches thick, but it was real hot, you catch on fire. Yeah, and go in and out. We worked in shifts to replace the bricks and it was real hard. That was hard work, I'll say.

My job title in Welded Tube was pipe threader. Well, I started at the Welded Tube as a crane hooker.

The best day in the mill was quittin' time. I worked three shifts, but I liked day shift the best. Oh, I don't know, it seemed like on night turn shift I never got enough rest or sleep, but I enjoyed work, though.

Well, my worst remembrance was when the galvanizer caught on fire. It was a big fire and everybody pitched in to help put it out. That was in the galvanizing department.

Yeah, well, there were a few men who I wasn't exactly crazy about working with, but I guess everybody has that. They were real nice, yeah, most of the men that I liked to work with passed away already.

Well, I'll tell you, I lost my hearing, all those years after about thirty some years they never issued no hard helmet. That was government regulation, hard helmet and ear plugs. It was in the 70's.

Well, we had three children. My daughter and son, they had a tour of the mill when they were young. They saw all this new, automatic machinery and they were watching me work. It's all automatic, and when I came home he said, Dad, you say you work hard, I didn't see you do nothing but put your hands in your pocket. The machine did all the work.

Oh yeah, they appreciate what I did later. Well, they're all up in New England now. My oldest son, he works for Draper Company. He worked on the space shuttle, then he graduated with a bachelor's degree from MIT up in Massachusetts, then he worked on the landing part of the space shuttle, and now he's on a different project. Now he makes three or four trips to San Diego, in fact, that's where he is at this week.

Yeah, if there was something I could change, I would like to go back to being eighteen, nineteen years old. Working again in the mill. Oh, I don't know, I just enjoyed that. My oldest son took a tour when they were in high school. He had a white shirt on like you, and he leaned on that railing and it was all dusty from all that dirt. He took one look at that shirt and said he'll never work in the steel mill like me.

My last day in the mill, well, they made it pretty easy. They let me . . . they put somebody in my place to work and all I did was say good-bye, then I went around talking to everybody. We had, like, a little brunch. Well, a lot of people don't believe it was hard work, we used to go through. The younger people today, they don't seem to be as ambitious like the older generation.

I know you don't remember the Depression or nothing. You know, when I tell somebody about the Depression, I went through the Depression with my parents, you know, it was tough, you know, and I used to work in a flower shop for five dollars a week. And I used to come home and give my money to my mother four dollars and fifty cents and I would keep the fifty cents for myself to go to the movies or you know.

Oh, that was way back in 1940, then I was in the CC Camp, too, for six months. The CC Camp was the Civilian Conservation Corp, begun by Franklin D. Roosevelt, you know. Well, I was down in Virginia CC Camp for six months and they wrote to me J&L was hiring so I came back. After six months you could sign up again for another six months. We use to make thirty dollars a month and in the Army used to make twenty-one dollars a month. They use to send twenty-five dollars home and we kept five dollars. That's when most of our people was on welfare, but you had to work for your money, not sit at home and collect. You know but it was rough going in those times.

Anthony Rivetti

His remembrance was polite and controlled in an objective appraisal of the livelihood the mill offered him. He sat with gray, wispy hair and a studious appearance afforded by his spectacles, almost apologetic.

My name is Anthony Rivetti and I'm from Aliquippa. I am seventy-one years old and was born on December 25, 1924 in Slovan, PA, a mile and a half south of Burgettstown Town.

I had thirty-four years, nine and a half months of service, and was hired on April 16, 1946. Yeah, right after the war. I was just twenty-one years old. I went into the service when I was eighteen, came out when I was twenty-one. I never had a job in the mill or nothing, cause back in them days it was hard to get a job. You know, even the older fellows up in their twenties, like my uncles, they were all loafing and everything on the streets.

Whenever I came back from the service, jobs were still hard to get, so J&L was advertising. They were advertising for men out here, so my sister and brother-in-law lived out here. So I went in, moved in with them, and I got a job right away cause they were, you know, short on men down there. They were really bad off. So I got hired right in, easy, real easy, didn't have no trouble at all. It was all right working back then.

Just the thing that I was really shocked at was when I walked in this mill how huge it looked. I looked at that building and the guard said, don't you know where to go. I said no. I said, they told me to get in that building, but I don't know how to get in it. He said, well don't go in the door by tracks, then he hollered at a guy to come over. One of the fellows that worked in the tube. He asked if he would take me down to the . . . down to where I was working.

He said all right and I was really amazed at all the pipe they had in that place. I wonder, man, who in the heck buys all this pipe, you know. I just couldn't believe it, even after working all those years down there. There was more and more and more pipe, you know, you know, still it was hard to believe. Who was actually buying all this here pipe? You know, but it was just like a good, good family down there. I mean, it was really nice working there. It was nice, it was something to do.

I only had the one job. I worked in the Inspection Department from the day I started to the day I retired. I inspected pipe, inspected half inch butt weld pipe. We had lappweld pipe down there. I went there, I looked at the continuous weld pipe when they got the new one in. The last mill they got down there was the Electric Weld Mill. So then I went there steady, but then, whenever the job was done. We had to go all over the place and work, you know. We also looked at couplings, too, but very, very seldom. I was on pipe all the time. I stayed in that Electric Weld Mill from almost the day I started to the day I retired.

These oil companies, and then we made real heavy wall pipe up to eight foot long. The piling used . . . to put piling in the ground, for bridges and roads and stuff like that. General Electric when I first started down there, they had a lot of conduit. General Electric and National Electric, both of them. That was all we were doing down there at first, maybe for the first five or six years I worked there, the General Electric conduit. You know, that's light wall pipe when they run the wires, and National Electric Pipe. Conduit was where they ran the electric wires. Wires up through it, like it was easy to bend and all. You know.

Best day, oh, I had a heck of a lot of good days. I never give anybody any trouble down there. Got along with bosses all real good. Didn't really have no trouble. Like I said, I didn't have any trouble with them at all. None of them ever gave me trouble. I always seem to have done my job and didn't get in no trouble. Didn't try to go hide or sneak off or nothing. I had work to do and I did. When I was looking at pipe I was actually looking at the pipe just like I was the one was goin' to buy. I mean I wouldn't want to sell you a bad piece of pipe. You know what

I mean, so I wouldn't want a bad piece either. So that was where I always was.

Really, I'll tell you the truth, really didn't have any bad days. I was easy to get along with. No, well I can always remember when they used to give prizes out down there, see, for safety awards and different things. They used to give different things. That was nice, what they were doin' over there. That was really, really nice. Yeah, later on, whenever they changed supervision, they quit doin' that.

Oh yeah, that was really, really good to give positive reinforcement. That was really good. I also meant, don't know whether I mentioned, but I used to work on galvanized pipe, too. In that job, they had to send you where you were needed, but there was times you stayed on one job. Like I say, conduit when I first started. That was a dirty job, with a lot of dust and stuff there, and the pipe was wet and all that. After I had a chance to move up whenever younger guys came in, well, then they put the younger guys on that job. You know, and that would give me the chance to rotate some, and the jobs I went on went from seven points to levee. I was practically eleven points steady all the time.

Well the only thing that really hurt it was like on holidays. We used to work Easter, we just worked. We didn't have no overtime for Easter, and I used to feel hurt when everyone else was going out on a picnic or going visiting, and I had to go to work four to midnight or something like that. You know, that was the only thing that hurt, all those holidays we were working down there. I had never, never worked a Christmas, as long as I worked down there. They used to always shut down for Christmas. I don't remember anybody working on Christmas.

Well, the last day I was on midnight turn, and the boss come around and said he was short a man or two and he was going to take me off the job altogether so I could walk around and see all my friends. Because I had to go to the Lasting Mill, to the Electric Weld, to the Galvanize, and then down below. You know what? I had a lot of friends because of jumping around like that. You get to know almost everybody on the jobs all over, whereas these guys working on the machines, those guys stayed on their job steady. They didn't have a chance to go from one place to the other where I was doing so, you know, I had a lot of friends.

Two of the guys, they were off and he said how about me trying to work you for four hours, would you mind. And I said, no, I don't mind. So I worked for the four hours and then they sent another man down to take the place. Then he let me go all around seeing my buddies, shaking hands, and then I got out of there. They had two of the prettiest cakes that you ever wanted to see, big, big cakes down at the Electric Weld.

I didn't even know those fellows were doing that for me, and they had like a party for me down there. They had coffee and cake, I mean, these cakes were huge and they didn't even finish all the one cake, and they wanted me to take it home, and I told them no, no, no. I said leave it here for daylight or something. They said, no, it's got your name on it, we left your name on it, take it home. So I ended up taking a piece of it home.

No, the only time is when you first started down there like in the Electric Weld Mill, there was an awful lot of noise. There is no way you are going to get around that, especially running the big pipe, you know. Take the three quarter inch pipe forty-eight, fifty foot long. My hearing, it's all right, now. What they did, one time all of us guys worked up there had to go down and get our ears checked, and they give us these plugs like to put in our ears. But after awhile there wasn't anybody wearin' them. You know.

The most important thing, however, was just to be very, very careful and do your job right. That's all you got to worry about, doing your job right. Like I said, I never had no trouble, none of them down there, not a one down there. I don't know, half of that I think was the Japanese that caused most of this. Then starting all these mills up across the pond didn't help us any. Didn't help us a bit.

Well, like in Europe when they open up all these different mills and stuff, you know yourself. When you buy clothing or steel, we're getting a lot of stuff from Taiwan, China, and those places. No, I just want to thank you for this interview, and I just hope I was all right for you.

Assorted welded tube elbows

He was a fiercely proud in his countenance and mannerisms, his suit and tie in sharp contrast to the aging union hall. However, he beamed with pride pointing out that James Downing, Sr. was a charter member of the USW 1211 Aliquippa Chapter. The black experience in the mill was different than the rest, more personal, and at times, I suspect, more painful.

My name is James Downing, Jr. and I'm from Aliquippa. I'm seventy-six years old and was born in North Carolina. Pitt County, Grifton was the town.

I worked forty-two years in the mill and was hired in 1940. I started making five dollars a day in 1940. I retired 1982, and was on piece-work then.

I worked in the Welded Tube. Well, back up a minute. See, I got hired in 1940 then you see, you got hired, you didn't get hired directly to any department. So they had, like, a labor reserve pool. So then you will go there, and let's say maybe they got five jobs available at the Byproducts. They'll send them five guys there and then they send ten guys to the Tin Mill, I mean like that. So I got, I worked in all the mill but the Byproducts and the . . . I never worked in the Wire Mill, I

Jim Downing, Jr.

worked in the Nail Mill, I worked in the Blooming Mill, 14 Inch Mill, and Open Hearth.

That was the labor pool then, I guess, I was in the labor pool. I'll say about a year, I believe. Yeah, right, see I went into the service in 1942, June 5th, 1942. Well, the thing is, what I have to say now . . . hey, you're black, you get all the bad jobs.

Well see, the union was in existence during that time, you understand. The union, in other words, to me is like I saw with General Motors when them guys were striking. They were striking due to the point that General Motors got someone outside doing the work. Therefore, they don't have to pay them outside as much as you pay the regulars. But anyway, what I'm saying is the union, it wasn't strong, understand. You know, and we called this the infant stage. A guy came up with a bright what not. The union committeeman, you know, sometime the committeeman he's in the pocket of the boss, understand, like that, but anyway . . .

I spearheaded some changes, you know what I mean, say 'cause the jobs. There were jobs that blacks had never worked on. Say you're working on this thread machine here, okay. I'm laboring, my job is regular labor and what not, doing labor work. But the thing is here I got, maybe I got five years in. Here is a guy, a white guy, he got maybe three years. Well okay, the foreman would put that guy with the three years in on your place there, not moving me up with enough seniority than this other guy, you understand. But so, these kind of things . . . but hey, man, I come up here many days and talked with the union president and the committeeman. Hey man, I'll pay my dues and that guy said I was entitled. I mean, hey, I don't want to step on nobody's toes, but I don't want you to step on my toes for you to get ahead of me.

My father, it's sad for me to think about, to this day. To this day, I don't know how we survived in this town of Aliquippa. Well, my father came here in 1923, my mother and my sister and I be coming in 1925. But anyway, as a child, you know you don't understand what, and back in the 30's this guy — I think his name was Al Anders — he was from Kansas. He ran for president and he wore this big sunflower, you know, big sunflower. I wasn't about to . . . I remember when Hoover and Smith ran. They were saying don't vote for Al Smith because he wants to bring back, what do you call it, you know, prohibition. He wanted to bring back alcohol or whiskey, like, you know.

So this guy, this Hoover got in, you know. So he got in, then things began to fall apart. Then Roosevelt got in and he took things, he really started putting people on, people that were doing WPA and NRA and CC Cap and all this type of stuff. But in the meantime my father was in the mission of getting democrats, getting the people to register, you understand. I'm going to tell you, to this day. I just can't understand how that we wouldn't run away from here.

You know, blacks — same thing apply to those of foreign extraction. Like I say, but I don't know. I still don't know. Then a lot of people was afraid to come to our house, because there were stool pigeons. They would call up and tell, you find in that book there — Mo, Kelley, all them guys, they were big shots. Policeman and what not and them. There was no such thing as a search warrant, no, no. They come in there and get you like the Gestapo, you understand, then, hey, they'll beat you up. If a guy would beat his wife up and the people would tell the cops. They come up there, man, give you a good goin'' over, yeah.

Oh, I'll tell you, this is around in the 30's. In the 30's, you know. That was the Coal and Iron police. Well, that's right. That's right. I see the time, there was a strike at the "Y", you remember the Woodlawn Hotel? There were state troopers, they came in on horses, they looked like young bats. They wailin' on people. I was standing right there in that hotel, I don't know if you remember or not at the corner there.

There was a door to go into there, right on the corner of Franklin Avenue and Station Street there. I was sitting up there while all this shit was goin' on out in the street and what not. Then there is a man that build next door, where Mr. Byrd live at.

They was Todd Mitchell and then there was another man, his name was French. He lived on Return Street. Those fellows were involved in the union. They took them guys up on Griffin Heights and they beat them guys up. I recall my father telling me at that time that the company bosses, whoever was down in the mill, told him. Jim said if you come get out of the union and move to management and all this kind of stuff we'll make you a boss, you know. That's like I'm trying to fly without wings, you know, at that time and no no he stayed in and helped form the union and what not.

So people would come to our house at night, and he would get these cards so he could write. He'd write his name down and sometimes he was out on the street somewhere and you give him a card or what not, and then they come by the house at night, or you could sit out on the street. Then if you meet them they give me the card. And I bring it home and then he brings it down the union. The Union Hall down there, way down there where the old Russian Club used to be down there. So by collecting those cards, that's how we enrolled people in the union.

Well, the union started to get strong, well, I mean stronger than when I got hired. I'm going to tell you, going back to when I got hired, my buddies, you know, neighbors and local boys on the street, they got hired in the mill. So that summer I was the supervisor at the pool over there, 'cause my father involved in politics.

But anyway then the Plan 11 pool closes in the fall. So anyway, I went down and put my application in down there at the employment office. See, my buddies and them guys got hired, you know. They hadn't call me, and so I told my dad, I said, "dad, you know, I haven't got no calls lately."

So he asked me, do you got your social security card?"

I say, "yes." So he said, okay. So I don't know, the next day or whatever it was. He must have went down there and saw somebody in the company about, I think about two days I got a call to come down there.

Now, one thing I found in life that if you're a man or a woman people respect you even if you may dislike them, you know. The thing is and I've heard, and I've heard from one man, this Mike guy. I was down at the Y, it used to be right down there at the corner there before they put the overhead bridge there. There were two Italian fellows had a big shoe shop there. I don't know if you remember that or not. Okay.

This black guy, he worked in there as a shoe shiner and he also worked in the mill. He and I got hired at the same time, but he had a family. I would stop in there sometimes and talk to him and what not. So I had just been over to the . . . what's it called. He gave me a card to go to work somewhere. Well, this white gentleman, he came in there, got on the stand, he would shine his shoes there.

I don't know what made him talking about my dad, he didn't know me, I never seen this guy. So he was saying, yeah, Jim Downing, he's just as much white inside as he is black on the outside. My father was black. He was darker than I am. So my ears popped up now man, cause he's talking: This guy, he's just as white on the inside as he is black on the outside, awful white inside.

He going on talking, after he finished talking I took this card up and I gave to him. He said, oh, you're Jim Downing's son. I said, yes, I am. This guy was named Roy Dye. He was like, I guess, assistant superintendent of Aliquippa Works, and this is the guy my father went to that I got hired.

So what I'm saying is that during the war I was overseas there and my wife, she wrote and told me that they are picketing at the old P&M Store up there on Plan 11. But they didn't have no blacks so they were picketing. She said, well, they're picketing the Pittsburgh Mercantile Store. You know, they don't have no blacks. Well, they didn't have no blacks, but we were all in the service. So the movement, they couldn't make no progress. There was the Aliquippa Civic League that was a black organization that controlled them.

Black, I mean Aliquippa blacks with these civic leagues, civic leagues. So anyway, the next letter my wife wrote me told me that my father and the union fellows, white and black, they were on the picket line up there. Hey, man, they hired this black young lady, she was a friend. We all went to school together. They hired her, so the picket was over, but that was when a lot of blacks saw the union then in action because the whites and the blacks picketed there. Before you had all blacks picketed there. You understand, and so that, it took the fear out of a lot of them blacks like that.

We were working together. They were brainwashed through the company, through other blacks, so to say. So anyway that open their eyes up, these blacks. So that we had power, and you could work together. That the union, the union, this thing of the white working along with the blacks. Okay, the company, see, which they seen before, you know, with the whites, the whites would always be on one side and just like when your parents or grandparents came here and what not, they were treated just like blacks were, you know.

But the thing is, so anyway, I recall there was a black family here and they I think I don't know if they were on welfare or something. Anyway this guy, Charles Laughner, he took to West Virginia and Pennsylvania state line and put 'em out and told him don't ever come back.

Now I was going to Franklin Junior High, you know where that was, don't you? Well, I got out of school, I'm coming down Franklin Avenue and there was a little white boy standing there selling, I think, the *Aliquippa Gazette* Paper, a little paper, it looked like this free paper that you throw away.

On the front page, Downing and Dempsey, they'll sue for libel. Well Mr. Dempsey and another man, his name was . . . I can't think right now, but it will come to me. But anyway, the thing is Mr. Dempsey, he worked in the mill down there. But he was a foreman down there that must've been back in the 20's, I guess. But they didn't want to give him no vacation and this man, this guy, he wanted to get back to his rights someday. This other guy and my father, them three, they were just like this. But the thing, like I said about this Gazette, they took my father and Mr. Dempsey to Beaver. My father-in-law — this is back in the 30's, I was about ten or eleven years old, you know, junior high. Well anyway, I married, my father-in-law, he had furnished their bond for my father and Mr. Dempsey to get out.

Okay, you heard Byrd Brown in Pittsburgh, you heard of him in Pittsburgh, lawyer. Well, you probably won't know, if you went to Pitt, I know you heard about them guys. Anyway, his father was a judge before he died, but this was back in the 30's now. They went to Pittsburgh and got Byrd Brown as a lawyer. I don't know the cost, but I know they settled it. But you see, there is so much history going on in my life and other peoples' life here in Aliquippa.

You know, just like I recall Plan 6, Plan 12, what's the street area . . . Franklin Avenue, no foreigners or no blacks lived in those areas. The only time that you were allowed up there unless you were working or doing something. But before dark, you got to get out of there. They didn't want us at the Plan 12 Pool, they didn't want the foreigners to swim in the pool. They said that you're too hairy, you had to go West Aliquippa to swim in the river down there. Whatever, all that kind of junk, all this kind of stuff.

It did get better. I mean, you see what you're doin'. What are you third generation? All right, then, you like my son and daughter, they are the third generation and they can go to college. So we benefit from your labor. Yes, your grandparents' labor, whatever, your father, what not.

Best day was when I retired, that was the best day. No, you see. As things progress through life, you know, things get better, things change, you know. It got so that we're in, if a job came up, the boss would come see. Jim, do you want that job, you know, for two weeks without. See, in other times, that wouldn't have happen, you understand. Like I say, and then in the closing of my time in the mill there. I recall this particular case when they had to hire women down there. There was two black girls, they got to fighting in the girls locker room. Well, fighting and stealing, you automatically fired.

So Bob Davis, he was the superintendent down there, so my boss at the time was Lewis Stevenson. I don't know if you know or not, but anyway, he's got some problems. But anyway, he came over there, where I was working at on the machinery there. So he said Jim, Bob Davis wants to see you in his office. Well, we had a helper, so the helper ran the machine, so I went over to the office there. Bob was sitting at his desk and I sat on the side of the desk. It was like you and I talking now.

Me and Bob, we had a relationship, in a way. He said, Jim, I got a little problem. I said, what is it Bob. He said, John Bell and this white guy. They got in a little confrontation there and they got to fighting and, you know, fighting, you know, they got in a fight, that is automatic suspension. I said, yeah, that's right. He said, can you talk to them? I said, yeah, I can talk to them. So I left there and I got my spell and then I went and talked to each one of them separate, and told them: You know that they told you stealing and fighting in the mill, you are automatic fired. He said, yep. I said, you know you could've been fired and Bob didn't want to fire you, and I told Bookem same thing. Never had no more problem. I get back to this black guy and the white guy, and they had an argument down on the job. So Bob, he sent for me again. I went over there and he told me, can you talk to John Bell, that's the black guy. I said, yeah I'll go talk to him. I said, what's the problem? He said, you told me that about the guys. I said, well Bob will tell you this now.

My wife and I have two children, a girl and a boy. When both of them are in a confrontation, we don't talk to only one, we talk to both of them. I said, now, you're in the position that you have to talk to both of them, you should talk to both of them. I said that. So this guy, the black guy, he had a high temper. Bob Davis was a good guy, nice guy, quiet.

I never heard the man speak out or raise their voice like some of the other rookies letting boss go to their head. He said, well, Jim, I don't want to be talked to with John. He might get smart or whatever, and then he is going to be fired. So I said, okay, I'll go talk to him, and I went and talked to John and I told him. I said, listen, John, do you know you could get fired for that. He said, yeah, I know that. I said, well, I'm going to tell you, Bob Davis asked that I come talk to you. I don't have to come and talk to you. You know, and I said, you can take it however you want to, so we got that resolved.

African-American steelworkers

The Plan 11 pool

Highspeed roller

His professional business-like exterior seemed adaptive to his job as an inspector, where significant financial gain or loss swung on his approval.

My name is Bob Jurasko, and I'm fifty-seven years old. I was born in Aliquippa in Sewickley Hospital. I actually worked, came out of there with — I have to think this over — 1959 to 1985 was after the war, so that's twenty-six years. It was 1959 when I went into the mill, I don't quite remember, don't think it was even two dollars an hour. But I don't remember to much about that because I came from a job that was paying forty dollars a week, and when I went there it was more money.

Bob Jurasko

My first department at J&L was the Welded Tube Inspection Department. I was an inspector. Welded Tube made pipe anywhere from half inch to four inch on the Continuous Weld (CW) Mills. Then I went onto our #1 mill which made half inch to one and a half inch on number 1 CW; two inch to four inch on number 2 and the Electric Weld made four inch to twelve inch.

The inspector would well, when they come off the mill the pipe would be made. Then it would go through straighteners and they would put faces on the end of the pipe and clean up the ends of the pipe. We would be on a bench, the pipe would roll over, we would look through the pipe for defects — the pipe seam with a coil in or bent over with a seam in. We looked for seams or any welds that didn't come together, so we had two inspectors, one on each side of the bench. You had a lead man and you had an inspector on the other side of the bench. We looked for defects in the pipes.

Oh yeah, there were quite a few defects. They still made pipe out of the scrap steel, so there were times when everything they made was scrap, but for certain specifications it didn't matter. There were low pressure lines where they could use steel not at the higher grade.

Well, basically the work was sort of irregular. The work was never that good, we worked a lot of four days, and there were a lot of down weeks because there weren't enough orders. The quality of the pipe was good, if they had good steel. If the steel was good they were using. Like I said, they couldn't make anything else out of it. So it was sent down the pipe mill and we made pipe out of it. A lot of that pipe was just used with steam, had no pressure on it. Basically it was the same all the way through. They made a good product to begin with, it really was a good product, when it was manufactured from good steel.

Welded tube was used for gas lines, waterlines, a lot of conduit, which was for electrical wiring, and that was our biggest customer. That's what happened when they lost the contracts with General Electric and so forth, they bought most of the conduit. They worked the weekends, the only times you worked weekends was when they were making conduit for electrical wiring. I guess just with the time plastic pipes starting coming in. I think a lot of that started happening.

Naturally when there was no work as an inspector, we had to labor. So you had to crane follower, a laborer. The crane follower just basically had to follow and lift the pipe where it had to go. I mean, you have to hook them up and had to take them down from one bench to another. Then put them on a basing machine or a threading machine, or so you would follow the lifts all the way down and you would unhook the cables, put the cables up, hook 'em up and follow 'em. Very dangerous work. Never thought about it as a young guy. I never thought about it, but as you start getting older you start seeing cable snap and you really start to think about it. At first it was nothing. It was just a job, that's all.

After being an inspector I ended up being in the office. I went into clerical and ended up being a order clerk. Then I ended up being a scheduler at the very tail end, about the last two years. I did a lot of scheduling in the labor pools.

I liked that a lot better than being a pipe inspector. When I was a pipe inspector I was out in the elements, the cold, the hot. When I went into the office, you always had air conditioning or heat. You didn't have to come to work in the same clothes and go home in the same clothes. It was a little bit easier. I even worked some of the shifts I worked on the half hour so I could drive in a lot easier in the parking lot because everybody was gone, I worked on the half hour.

How did I get a job like that? To tell you the truth, in 1976, I ended up with a back operation. I was home for six months and didn't know if I could go back to do the job as an inspector, cause an inspector was always bending over. As a matter of fact, most people ended up with bad backs. I called the mill and they told me there was a clerical job opened. If I wanted it, they would hold it for me, and I took it.

The best day in the mill, I think the best days were when I actually got steady daylight, I'd have to say that. When I finally didn't have to work any more shifts, I worked from eight-thirty to four-thirty. Yeah, when they told me I was going to go on that, no more shifts, I thought that was the best day I ever had. I could finally see my kids again cause all those four to midnights and midnights, it was tough when you're raising a family.

Yeah, and tough on your family and you, 'cause you never see some of the occasions they're going through. Couldn't be there for some of their plays or some of the band concerts or some of this, 'cause you had to go to work.

J&L was one of the few places. There was a lot of steel mills where you could trade for one day with somebody. Not our place. You had to trade for the whole week, if you needed an hour off on four to midnight you had to trade with somebody to take the whole week.

I don't know, I don't understand why they never wavered on it so you could waive your overtime, because they did it in other plants. I know Armstrong did it and I thought that was great. Cause if you really had something to do all you had to do was have one guy trade for you for that one eight hour shift, and then you go back on your regular shift the next day without any overtime. You signed a waiver, I always thought. I think that was the toughest part, not seeing your kids, all the good times of their lives, proud times. The worst day was I was an inspector and we were standing around a fire. It must of been probably a coke jack, and everybody was standing around this coke jack trying to keep warm. It must've been in the teens outside. We were working in a wide open space that was cold, and it just so happened that I was running, I was the lead inspector that time. When the bed went to put the pipe down, I'd have to be the first person to touch this pipe. I walked over from my position at the coke jack over to the bench, pulled the stop out of the pipe, and the guy on the other side came over to do his job.

In the meantime, the man that was running the coupling machine to put the couplings on it, which threads automatically, walked over. I started to walk over, leaned over, hit the button on the coupling machine and the hydraulic line busted. The hydraulic fluid shot back at this coke jack and turned into an inferno, which caught this guy on fire as he was running away. He was engulfed in flames and everybody was trying to chase him down with fire extinguishers, and he was trying to roll over. It so happens that this gentleman died, and it was probably the worst day I ever seen in the mill because he did die.

Did it happen a lot, things like that? Yeah, I've seen guys have their fingers cut off. You know, didn't even realize it, the glove was still on their hand yet — hands smashed, lifts break, fall down and just miss people, five hundred half-inch pipe fall down, and just miss everybody. I wonder how I ever stayed alive.

I was in the mill and was ready to leave the mill and was washing up. I told my mother I would meet her at Gem Jewelers because we were going to buy my step-dad a radio for his birthday. I can remember pretty happy, pretty happy, and we walked down from Gem Jewelers and he put on his radio so we could listen to it, and it just came across that President Kennedy was shot. That one sticks in my mind because I knew exactly where I was and was so happy about getting there to get this radio for him, you know. Yeah, I would probably like to forget that fire. That fire, that's something that stayed in my mind for a long time.

I couldn't get that out of my mind for a long time because I thought to myself, I was the guy that actually initiated the move where everybody moved away from the explosion, and I have not moved so fast many more people would have been in the line of that fire, but I just went over to do my job and people left with me and that was something.

When people mention his name — cause there was people in my family that knew him and when they mention his name at a family gathering sometimes . . . Yeah, they'll say weren't you there, Bob, or something like that. It's tough, I like it to be out of my mind as much as I can, but it doesn't.

I would say a boss had the most influence on me. I would say one of my bosses, he was a foreman by the name of Paul Barr, and I think he was. Paul basically told me like it was. When I got there, he told me to get the hell out of here, don't stay around, but if you're going to stay here, make the best out of it, learn every job, which did pay off for me. Because it made it a little bit easier because I did get some of the easier jobs at times. I could do them when somebody was off sick or on vacation, and I think he did have an influence on me somewhat.

I had to work in the Galvanized Department and we had a little shanty over there. We were testing a galvanizing fluid, so we had go over to the acid tank and we had to bring the acid back into the lab. We put it in tubes and then we put different solutions into them, different chemicals. If they turned a certain color, you had to go out and tell the boss and they would have to dump the acid tanks and put a whole new batch of acid in there for them to get the pipe in before they clean it and it goes through galvanize. Well, unbeknownst to me, if it ever turned pink, the color of the solution, that meant that the tubs were bad, so you had to go tell the boss immediately. But it should never be that pink because someone should've checked it on the other turn. How could they not see that, but, well, I left the office, went out, came back and started doing mine, putting my chemicals inside the acid. And it turned pink so fast I couldn't believe it.

I said, what the hell is happening here. Well, it just so happens that I found out later my buddy had made a batch of Koolaid, cherry Koolaid, and filled up the solution and mixed it with the Koolaid as I was dipping and dropping it in. I didn't know that. The boss said you're full of baloney, there is no way this acid can be bad. We just changed that on the four to midnight shift today, but them guys put that Koolaid in there and, to this day, they say, hey Jurasko, you drunkin' any Koolaid?

Sad story is that by being a scheduler, I had to put all my buddies names on that list to be laid off 'cause they would tell me how many they were laying off, and we figured out by seniority. I think I went through this for about a five month period before we actually shut down. I put their names on that list, it sort of took me back, writing my own friends names down. Some people, it didn't bother me 'cause they weren't that close to me, but my own friends . . . There were some days when I wanted to put my own name on that list. We knew that we were going down for probably . . . I can't say. Well, we started January of 1985, that's when we really started. Every week a different group of guys would go, a certain job would shut down and they combine two jobs.

Little by little we started phasing out. So every week there was a group of guys who were getting laid off. I worked all the way till March 23rd of 1985, and at that particular point in time there was hardly anybody left in the plant and in our department at all. So it was tough, it was hard seeing my buddies nervous. This is forever for them.

My last day in the mill I went to work that morning and asked the boss, am I working next week. 'Cause I got to make my schedule 'cause there is only a few people left. Am I working, cause I would work until the end. I would work until the last guy worked. He said, Bob, this is it, we're pulling the plug next week so, he said, that's it. So I went to work that day, you gathered up your things you were going to take home, things you had left in the mill for all those years. For twenty-six years they were sitting in the mill, in my locker, or in my desk.

Now I had to gather all those things up and put them into a shopping bag, wondering if I should throw them away or should I keep it — should I take schedules home, peoples names, phone numbers, for some reason if they should ever start the place back up, you know, just thinking maybe that would happen. I was sort of gathering my thoughts should I do that, should I not, but it was a long day, that eight hours was a very long day. I mean, there was nobody really around except me.

No, I really didn't think that it would start up again. But then later, a few months after the place was shut down, they did have some prospective buyers and they said they would contact me for that particular list I was telling you about that had phone numbers, and could I gather enough people who would be interested in coming back to work, and they did start back up, you know, and nothing ever resulted from that.

It was tough on shifts because my wife at that time was home with the kids. I had two daughters, and as the daughters grew up, my wife she decided to work part-time but she didn't like it. She didn't like midnight because she had to do things around the house. She tried letting me sleep but that sort of cramped her style. I had to tell the kids to keep quiet as the kids got older. If it was four to midnight she wouldn't see me.

I would come home from work, she was sleeping, and then later on down the road when she started to work a little bit, it was worse yet because I would come home from midnight and she was gone. She come home from work at four and I say I'm going to go lay back down again. You know, it wasn't a good life. I wonder how I ever did it, in the cold, in the heat, but I guess I did it cause I had a family to feed.

No, I don't think any of this could be changed. I think there was enough blame down in the steel mills — the management and the labor. I don't think I could have, I feel that I did the best that I could do to keep the place goin'. I did the job the way I thought it should be done, and if that wasn't good enough, then I don't think they had any intentions of keeping it.

Management did — it depends on who the management was — some guys. The way I always looked at management was like this, those guys put their pants on the same way I do. I was never one of those guys that said, well, the hell with him because he isn't a boss, or, I would never go out in public with him, or, I'm not drinking with that guy, or, if I ever see him I'm goin' to punch him in the mouth. That guy had a job to do just like I had a job to do, and some guys, they turned out to be pretty hard on you and they had guys on their backs. They had somebody above them. A lot of times it was somebody who had just come in from someplace else. They knew nothing about our department and how it was ran. The rank and file, you had your good and your bad just like any thing else. You hear people talk about how they sleep down the mill and that was the reason for the demise, and those that robbed the company blind — you can't stereotype every steelworker like that. I put myself into a lot of my friends' positions.

We were there everytime they called us. They want us to work, we were there. You know, you could have guys who never wanted to work a Friday night, they would report off, and they were probably guys who did sleep in the steel mill, or guys that stole from the steel mill, but don't stereotype every steel worker that way, you know. It's not my fault, and I say this to people, and I'll say it again here, I blame it on the government and local government and state of Pennsylvania. I see LTV mills still goin' strong in the state of Ohio, Cleveland, and out at Indiana. So the way I look at is if we had no help whatsoever I remember when Crucible Steel in Midland was shutting down, they called Governor Thornburg in to put his nose in down there. He wouldn't even show up in Midland. We had no help as far as coming down here. I think they could've survived in Pennsylvania, if they would of gave breaks like Ohio did. They have to be giving people breaks to survive out there. Why did they go to Ohio, they could've stayed right here in Pennsylvania if they had the same breaks. I think what it came down to was dollars and cents.

One thing I will tell the young, and I'll continue to tell them to my dying day is don't ever tell me the place is like it is because the union did that. If it wasn't for the union, people like you in an office job or somethin' other than steel mills are making nice wages because the union got that. They weren't brought up to them kind of standards without the union helping out. I tell it to my buddies in US Air all the time.

They're in management. They always tell me, yeah, those union guys, they're goin' to break this airline like they did the mill. I say how much do you take home every payday, you think you'd be taking home that much every payday if there wasn't a union? They got to pay you more than what they pay that man out there, right? So that's what I tell the young guys, don't knock the union 'cause, as a matter of fact, someday they may be asking for a union to come into work.

The only thing I would like to say is I think the forgotten thing in this shutting down the steel mills is people, the young retirees. Everytime you talk, you know everybody talks about people on Medicare or they talk about people who have. I was in the prime of my life and I had to go back to school. I went out and got myself another job — not by choice but I had to do it. I think the younger generation, the younger retirees, they are out there and as a matter of fact, I'm the youngest officer in the Steelworker Organization of Active Retirees.

When I go to conventions I get told by people at those conventions. What the hell you doin' here, you're still wet behind the ears? What the hell do you know about the union? And I get tired of hearing it. My wife heard that and almost cried and walked away. They're telling me I didn't march on the lines in 1936 and 1937 with the guys getting their heads beat in. For young guys like me who try to stay active in some of these things, as the older ones begin to die off . . . There aren't goin' to be too many people to keep these organizations going. They should be out here trying to find some way to get people interested — retirees, organizations, and so forth in my age group.

I'm only, like fifty-seven, and a lot guys are younger than me that were retiring, but nobody seems to care about those people out there. Everyone is talking about and worried about Medicare, which is a big thing. I'm for everything they're talking about, but they have to find a way to address some of these people that haven't made it back yet at age fifty-six but that haven't found a job since they left the mill. And if they did find a job they're sweeping a floor someplace for three or four dollars an hour.

Welded Tube inspection

Welded Tube laboratory

Former Welded Tube sites

Wartime effort

ALQUIPPA WORKS
WELDED TUBE MILLS

Production Data

Annual Capacities and Product Sizes

Unit	Size Range of Product	Size Skelp Used	Annual Capacity
A-1 Continuous Weld Mill	½" to 2", 17' to 51')	.098" to .300")	420,000
A-2 Continous Weld Mill	1-1/4" to 4", 17' to 51')	6-5/8" to 17-½")	
A-1 Electric Weld Mill	5-9/16" to 12-3/4", 16' to 55'	.188" to .375" 20" to 43"	216,000

Equipment Data

Unit	No. Of Units	Description	Size Of Motor	Speed
Coil Slitter	1	42" x .098" - .276"	125 HP	100/300 FFM
A-1 Cont. Weld Mill	1	1 Uncoiler, Mandrel Type		
		1 Roller Leveler 20" x .3125" Maximum	75 HP	304/1215 FFM
		1 Double Upcut Shear	75 HP	522 RPM
		1 Flash Welder, 500 KVA		
		1 Preheat Furnace 102'-11-1/4"		1600° F.
		1 Heating Furnace 154'-5"		
		14 Stand Forming, Welding, Reducing Mill	30-40 HP	1250 RPM
		1 Rotary Flying Hot Sew 17'-51' Pipe Lengths		1250 RPM
		3 Stand Sizing Mill	40 HP	1400 RPM
		1 Rotating Screw Cooling Bed 38'		
		1 Splitting Saw	50 HP	1750 RPM
		2 Crop Saws	50 HP ea.	1750 RPM
		1 Water Bosh Cooling 58' Misc. Finishing Equipment		

Reference 2

WELDED TUBE MILLS (Contd.)

Unit	No. Of Units	Description	Size Of Motor	Speed
A-2 Cont.				
Weld Mill	1	1 Uncoiler, Mandrel Type		
		1 Roller Leveler 20" x .337"		
		Maximum		225/900 RPM
		1 Double Upcut Shear	75 HP	530 RPM
		1 Flahs Welder, 750 EVA		
		1 heating Furnace 190'-9"		
		10 Stand Forming,		
		Welding & Reducing Mill	50 HP	700 RPM
		1 Rotary Flying Hot Saw		
		17'-51' Pipe Lengths		700 RPM
	1	3 Stand Sizing Mill	3-50 HP	800 RPM
	1	Rotating Screw Cooling Bed		
		37'-11"		
	1	Splitting Saw	100 HP	1750 RPM
	2	Crop Saws	50 HP ea.	1800 RPM
	1	Water Bosh Cooling 58'		
		Misc. Finishing Equipment		
A-1 Electric				
Weld Mill	1	1 Uncoiler		
		1 - 7 Roll Leveler	50 HP	1150/2200 RPM
		1 Upcut Shear		
		1 Side Trimmer	50 HP	1150/2200 RPM
		1 Scrap Chopper	25 HP	875 RPM
		1 Vacu Blast Unit		
		1 Forming Mill - 9 Driven Stands	(7-30 HP	1150/2600 RPM
		- 1 Idler Stand	(2-50 HP	1150/2600 RPM
		- 4 Edler Vertical Stands		
		1 Rotary Transformer Welding Unit		
		2000 KVA - 250,000 Amp.		
		Electrode Diameter 66"-39.5"		
		2 Pair Pull Out Rolls	50 HP	1150/2600 RPM
		2 Amealing Units		
		3 Stand Sizing Mill	2-30 HP	1150/2200 RPM
		1 Flying Cut-off, 15' Travel		
		- Forward		150 RPM
		- Return		180 RPM
		Miscellaneous Finishing and		
		Testing Equipment		

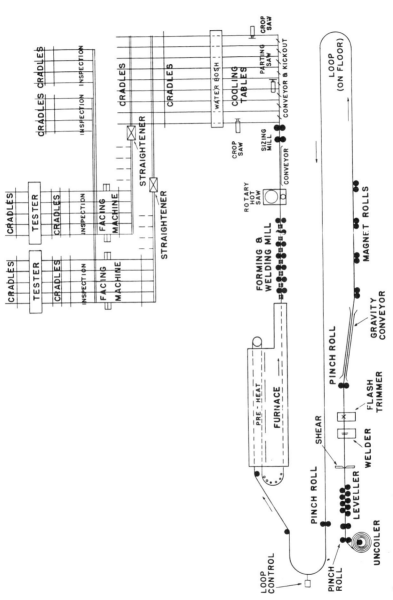

SCHEMATIC SKETCH OF A CONTINUOUS WELD PIPE MILL

Reference 2

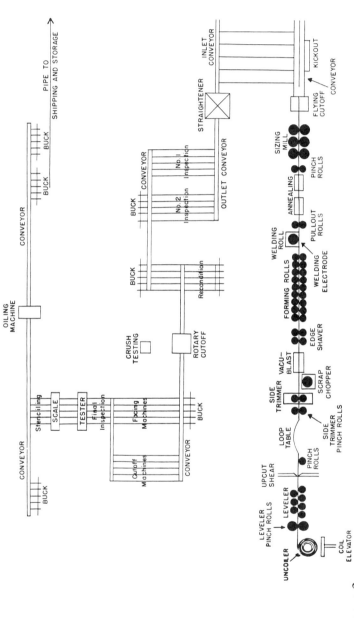

SCHEMATIC SKETCH OF AN ELECTRICWELD PIPE MILL

Reference 2

Chapter 10

ℰ✺Ↄ

Seamless Tube

H e had a distinguished appearance — gray swept back hair and a dense Italian accent — seemingly articulate and well-educated. Like a detective, he presented his case for the rise and fall of J&L.

My name is Louis Salvoldi and I'm sixty-seven years old. I was born in Europe, in Italy, and was hired at J&L in 1947. I worked for thirty-eight years and my wage was one dollar an hour. When I retired, I made twelve dollars an hour plus bonus. Oh boy, I know it was 1987, something like that, 1986 I think.

The Seamless Department, 30 Inch Round Mill is where I worked. So I worked in the Seamless Tube and 30 Inch Round Mill. We were making rounds, well, in the Seamless Round Mill we washed the steel off, we get these ingots called blooms from Blooming Mill. You know why its named bloom? It was a long ingot like a railroad tie and then we heat it up, then we rolled it, it was squared and we make it round. It would be like, maybe fifty feet long round, okay? There were two of us, we would feed it back and forth on the roller, right, like over here. You go back and forth reducing to right thickness. We were supposed to cut it down the way it suppose to be, and then go to the other mill to finish it. You go and finish, and it come out like oval, and then go to the other place it would come out round.

Louis Salvoldi

The last year I was doing this job called table operator. This mill installed in 1927 was really long, the tables were fifty feet long. They would be carrying steel, taking it one person to another. The motor of that was 3,000 horsepower. It goes forward and reverse. It went forward and reverse until the bloom was shaped into a round. It would go really slow and then pretty fast.

I did a little bit of everything. We would load these rounds take them over to seamless. I worked in the furnace pushing the rounds out from the furnace. I worked in the cutter and worked in the back of the furnace. All these jobs were hot.

You start from the bottom and move up, so you did a little bit of everything. My best day in the mill, do I remember a really good day? Well, especially, in the end was good, making good money. I always worked a lot of overtime and was easy to pay the bills. I was pretty happy, pretty happy. There was a lot of overtime.

Well sometimes — I think it's politics — one time they wasn't drilling anymore oil wells and didn't need pipe. The seamless tube, they didn't have no jobs, cause there wasn't a lot of demand in pipe. But later in the year, there was a big demand in pipe and the United States, not the other countries, could provide enough pipe to this company so we have to work. All the oil companies used the pipe. They used most of this pipe for drilling. Then the oil stopped, and then the stock was used. Then, they used brand new seamless made in Europe, about four to five miles from where I was born.

Well, a bad day was like if you ran the machine, a very hard machine to learn. It was a really hard machine to operate. When you go up there, if you have trouble, forget it. Forget it, 'cause if you think about the trouble each of those pieces of steel there may be 10,000 dollars or more. There be times you thought how much it was costing. No, I never, well I'll tell you what, if you do the job and you say it don't matter, you're going to get hell for it anyway, no matter what you do, you're going to get hell.

No matter how you do, you'll never get it right, somebody would tell you. You know, put every piece of wire over here, some there. It was very, very complicated. Everyone would tell you don't do it this way do it the other. Well, right now I was dreaming last night I ran the machine, and it was in trouble. Even to this day I remember the trouble. I don't even do it anymore. Even my buddy, he called me a couple years ago. He said, you still think you can run this machine? I said, I don't know.

Well, like I said, come from old country, you can't read or write, can't do the English especially. You go get a job in this country, good paying job, what more could you want. If I had a good education it would be something different, but like I say, even if you got a good education and you got a job, you may be making big money. There has to be something there for the trouble, big money, but it's a lot of headache, too. So like, there you got a pick and shovel you were making a little less, nothing to worry about, but it was hard. It wasn't air condition, but don't worry about getting hell cause you're always going to get hell no matter what you do.

My boss say one time, he said I don't want you to run any more cold steel if you roll any more cold steel, I'll fire you. Then, twenty minutes later, I told him why you send it if it's cold, don't give it to me. So twenty minutes later they went ahead and gave me cold steel, so I call him out, he said, go ahead and roll it. He told me to roll it. It was all the job, and when you were in this job, you had to make up your mind, and put your head in it. You had to cut the steel evenly, but didn't have enough time to do it. And you had to do your best or else. And watch don't break the saw one mistake and you break the saw.

My wife know that I had a job to do, and that's that. There was a guy who would tell his wife everything. I never go home and tell my wife what goes on in the mill. Never, just I had a job to do, I got problems, I'm busy so what, but well, today you got troubles, tomorrow you start another day. Well, I never really had any trouble, you know. Usually even the boss say I give you trouble, but you never answer back.

Alex Dudak

Well, that mill, they scrap it because it was too old. Well, the new mill nowadays have these round mills because they used a continuous Strandcaster. That's right. There was Blooming Mill audit, there was maybe a bunch about 100-125 people in the Round Mill, and maybe another 150-200 people in the Flat Mill. Now they all be gone and everything.

You know, some people say this is a good thing, times change. You got to go with the progress. What they did over here to the mill, I'm sorry to live to see it. There really isn't much to say. I'll tell the young people to go to school, be smart, 'cause today if you don't, you aren't going anywhere in America.

Seamless department jubilee parade

Seamless floor rehabilitation

He was a large man — a huge hulking presence, but I was most impressed with the gentleness of his nature and manner, almost quiet in his speech.

My name is Alex Dudak. I'm seventy-one years old and was born in Aliquippa. My folks came from Czechoslovakia in Europe. I worked for thirty-two years. Well, I came out of the service, went in 1946. I worked for about three years then went to school, Indiana State, but it didn't work out. So I came back in 1950 and then got a job at J&L. I worked until '82 — thirty-two years. Well, see, I had surgery, had to take a medical discharge, you know. Well, I worked in the Seamless Hot Mill, and worked hard for a living.

In the Seamless Tube Department we made pipe, you know, it was a round billet without a hole and we put a hole in it, different sizes, up to fourteen inches — three-eighths of an inch to a fourteen inch. The pipe was mainly used mostly for oil wells, different sizes, the thickness of the pipe, you know, some thickness down to about an inch. It was hard work, all hot, dirty. You made good pipe then. We made pretty good money in the Hot Mill, too. It all depends on how many pipe you put out. Yeah, our pay was dependent on how much production you had.

About five years before retirement I went into the Inspection Department. That was the best job I ever had in there. It was a good job. In the Inspection Department, we inspected pipe. I didn't do nothing, you know, let 'em go. There was a pipe one guy on each end and then they eliminated one man after layoffs. Well how the hell are you going to look at both ends with one man, but it was a good job. I'll tell you, I liked the inspection, you know. So you were to pull out the bad pipes. It all depends on the production run. Sometimes it would be five or ten pipe or maybe twenty or thirty in a lift that were bad, you know.

But I did most of my time in the Hot Mill, I made that pipe. The pipe in the Hot Mill Furnace. First you put a hole in the pipe, that was the first operation. And second operation made the hole a little bit bigger, and the third operation you finished it off. There were about five operations before it was down on the cooling rack where it cooled. It was hard work I'll tell ya. I don't know how I did it — hot and hard.

You couldn't stand it in the summer. There were big fans, I use to have to wear winter underwear otherwise I would get burnt. I would be changing clothes, its ninety degrees and I have to put winter underwear on, tops and bottoms, to keep from getting burnt. Yeah. I don't know how I did it. Hot and hard. A lot of heavy lifting, plugging the mill on the furnace, all the jobs were bad. We had guys come in one day and they quit, they wouldn't work in there. I don't know how I did it, I had a family, you know. The inspection days, yeah, they were good days, but none in the Hot Mill. In the Hot Mill they were all bad. It's a tough production. I worked that for a good number of years, well, about twenty-seven I'd say. So, they were all bad days, all of them were bad.

The dirt, we turned coal dust into pipe, to lubricate it. And coal dust, smoke, you couldn't see a guy, we were in that smoke. I think that affected your health. Well, Tom Morgan — I don't know if you know him — he got cancer, I think. A lot of guys, most of the guys are gone. Most of the guys are gone, and you think its somehow related to J&L right. I got hit with a crane. You know how short this is a pipe rolled down on, it was on a slant, you know. You had to roll them, then you get them down to the bottom, then we pull it out. It was hard work. I hate to even think about it. All those jobs in the Hot Mill, I would like to forget every one of them. There wasn't a good job in there. It was like a dream, you know, I was in New Guinea and the Philippines in World War II and it was hotter in that Hot Mill than in New Guinea.

Tom Morgan is a fellow I remember. He moved to Florida, though. He plugged the High Mill for about ten years and that's hard. There's a fan blowing on you all the time. Fans all over the place, and the noise, there was so much noise, I got a hearing aid. We had signals, whistles, that's how we talked to people.

We had a guy who plugged the High Mill, it's like a pair of scissors. You had to go in there and get that plug and the guys slipped and the pair of scissors went through his mouth and killed him. A lot of furnace guys died, from heart attacks, all the hard work.

What effect did this have on my family? Well, my wife passed away, but I can't blame that cause I worked in the Hot Mill. Well, I come home and I be so tired, sometimes I couldn't even wash my hands cause my arms would be so tired, sore. In certain ways it affected your home life, take it out on the kids, you couldn't sleep, especially on midnight, you couldn't get no sleep, and you had to go to work at midnight. Like today, it would be ninety degrees and I didn't have air condition in the house then. Shift work was hard daylight, four to midnight, and midnight.

If there was something you could change with the mill, no, it should remain the same. I'm glad it's shut down. Yeah, I didn't want my kids to work in there like I did. My kid is in the service now. He got thirteen years in, he goin' to stay until twenty and the other one owns a business. I'm glad they shut it down, nobody should work like we did down there. Of course, I made a good living, you know, but worked hard for it.

Well, in a way it was good, cause if you couldn't get nothing else, at least there was J&L, it was a job. But nowadays there's no jobs, flipping hamburgers at McDonald's and Burger King.

The effect on the community was staggering. Everybody worried about next payday. Of course, I went on pension, they shut us down, and I get a pension for J&L and I get social security, so I'm okay. A lot of them younger guys didn't have the time. They only have so much money, they got terminated. A lot of them didn't get nothing. A lot of them were laborers or just fill-ins. Did they stay around here or did they leave? Oh, I don't know where those kids went. Yeah, I know one kid bought homes, boats. They bought new cars and I'm telling these guys they're crazy. They want twenty turns. Seventy-five they start working real good. You work everyday, except Sunday daylight.

See there's twenty-one turns in a week, so when they say twenty turns that means Sunday daylight, that is the only time they don't work.

J&L was all right after we got a union. I would of never got a pension if it wasn't for my union. The union was a good thing. It was the best thing that ever happened.

My dad worked forty-nine years in J&L, and I remember he use to go down to work and I used to take a lunch. He didn't know if he was going to work that day. I would take a lunch maybe at ten in the morning, and have to leave it with the policeman. My father's name was Adam. He worked many years, probably forty-nine. He worked in the Wire Mill, in the galvanizer part. He worked like a mule, I tell ya. Well he got a pension, I think 150 dollars a month. He's dead now. He retired in the late fifties, I think. Oh yeah, in the Wire Mill, I know they have little cranes. Before, they picked everything up by hand. They didn't ask for a union because they used to treat the men like slaves. If the boss didn't like you, they fired you. You couldn't go to the bathroom. If he liked another guy he give him five days and give you two days of work.

I was just a little kid when the union started, I remember, about 1930, '31, '32 something near that. I remember the big picket lines down there. It was a big strike, you know, where the Y near the mill entrance is at. There were thousands of people down there. I remember this mill truck trying to get in there and they upset it. Better everything — work, wages, and safety. We had a lot of strikes after that, you know, for more money.

That was the Wagner Act, you know when the Wagner Act became law. The right to bargain with the company. There was a big strike down there, I remember the whole Y area of Aliquippa near the main gate was filled with men, because the guys that were scabbin' in there didn't have nothing to eat, so they tried to send a meal truck in there with food, but they upset that thing. I don't remember, I was just a kid, I must of been what, ten years old or something. I remember my father taking me down there though.

Like I say, I wish I'd have kept on in school. I wanted to be a school teacher, you know. I met a girl, bought a car, and hey, how are you going to go to school, you know what I mean, the government paid for the schooling, but still. You get a car and you want to get married. So I got a job at J&L. Anybody in Aliquippa went to J&L, that was the best job around. It wasn't a good job, it was a living, you know.

Tractor with seamless pipe

Seamless casting unit

He was a big man, large features with skin redundant with age changed from the muscle and strength of youth lost.

My name is John Kennedy and I'm from Coraopolis, Pennsylvania. I'll be seventy-nine shortly. I was born in a little town called Groveton, and that is in Coraopolis, on the outskirts.

I worked for thirty-three and a half years. Began in 1946, I think it was or '47. When I was hired I made eighty-nine cents an hour. I retired in 1980, and was making around fifteen dollars an hour.

The Electric Department and the Seamless, I was a motor inspector. Well, in general we made the seamless pipe. They took a ingot and they run it into the Round Mill and made it round in different sizes. Then they were all cut off four foot, six foot, whatever it was, and they came to the Seamless. They were preheated again and then they went through the Seamless Mill, which is the process of rolls and piercing points. They just pierced a hole in the seamless tube. My job was to keep cranes running, motors running, and all the electrical repairs. I was only one of about forty men.

I'll tell you about a day that was almost tragic for me. We went out to work at eight and about eleven a fourteen inch air line blew up and

John Kennedy

ruptured. So the pipefitters were called in and the Bull Gang and that to make the adjustments to it. So they put on what they called a Dresser coupling. That is a coupling that has two ends on it and is pulled together with bolts nice and tight.

Well, before they had gotten it apart they sent myself and Frank Ambosi, another motor inspector, they sent us up to cover all overhead wires, because there was 440 and 250 volts and 25 cycle all running through. Well, Frank and I were up there doing that, the pipefitter said close the Dressing coupling. Somebody signaled to Joe Burns to turn the air on, so when he turned the air on the pipe split back open. I was directly above it and it sent me fifteen feet up onto the catwalk. I landed on the catwalk and you couldn't see nobody for dust and dirt.

Now Frank Ambosi, he went off the side of the crane where the collectors shoes are and the collector bar. As the trolley moved back it shoved Frank to the end of it, and he received a broken arm or something out of it, but I had a broken back. See, when I went flying in the air something told me to grab the hand rail and I grab the hand rail and I flipped over and landed on my back, and I broke my back.

Well, I was there for a good fifteen, twenty minutes, half an hour and nobody knew where I was at, and I wasn't too conscious of what was going on. So finally, I reached over and got my flashlight out of the seat and I lit it and I started circling it, and somebody down below said there he is, up there, and that was me. Well, they came up to get me and they carried me in a blanket stretcher from there down onto the crane. They passed me from man to man because of my back, and then they brought me down the steps and they took me to the dressing station. The doctor up there said there was nothing was wrong, everything was good.

I lost a year's work because I had to have my back operated on where it was broke. Michael Zernich did the fusion of the spine, the last five lumbar. Now that was a tragic day of my life. I lost a whole year's work and I never got compensated for that, even to this day I've never been compensated by J&L. So that's my story I have to tell.

Well, I'll tell ya, the best day was when we could pull a joke on somebody. We had this one foreman, who stuttered. So he brought his lunch in a pail and we drilled a hole through the bottom of it and put a bolt in it and bolted it to the steel table. That was Wayne Swerensen. When he went to get the bucket it wouldn't come. I never heard the man swear or cuss or anything until that day. Then another day, Ben Holmes, he was our immediate superior, he was out standing by a barrel, talking. Somebody snuck up behind him and painted both shoe toes red, and then they got away and when Ben seen that he flipped his lid.

I remember us going on strike, it seemed like everytime you turned around down there, the Seamless, they were going on strike about something. Well, this one time when a fellow was off sick and I had to run a crane I was up running the Hot Mill crane and everybody was motioning to come down, come down. So I came down and they said what are you doing down here, and I said they told me to come down. They said everybody is on strike, we're going home, so they said get yourself up there we got to empty these furnaces. So I stayed there and I emptied the furnaces for them but when I came down I said may I go home. Because I was scared, I didn't know what would happen to me when I got out of that tunnel, you know, because I stayed and did my job and that was it.

After the war, 1946, there were a lot of bad relationships in the Seamless. We had a supervisor — I won't mention his name — he was an alcoholic and at that time alcohol wasn't considered a sickness. But he used to say there was no excuse for a man to get sick, but for a man to get drunk there was an excuse. We used to always laugh at him. That was his slogan, nobody get sick, but it was all right to get drunk.

No, I'm not a college man, but I'll tell you, we brought in a lot of college people and we had to teach the college people the mill stuff. Well, I'll tell ya, this one day they sent over a motor inspector apprentice. He had more tools on him then Sears and Roebuck carries. He said, they sent me over here to help you, and I looked at him and I told him, hey, when I have trouble I don't want Sears and Roebuck salesman coming, I want the boss. I said, you go back and tell Ben to come over here, I said, when Kennedy has trouble, the Seamless has trouble.

Well, I had a heart attack in February of 1980 and when I came back to work, why, Dr. Tauber wouldn't pass me and Dr. Sia, my doctor, wouldn't pass me. I said, how about can I get a disability. I said, you can't let me work and I think I'm able to work. But they said look at it this way, you're over sixty-two, it's time for you retire. In other words, they just shoved me out.

Not too much of my health, but it did affect my nerves though. That is why I'm hard of hearing today is because of that ring of the pipe. That pipe would roll down in these bucks that they have you know and it would bang. It could wake up the dead.

I would tell them to do like I did. Go in and don't know nothing, and then come out with a top job. That's what I did. I knew a little bit about electric, how to tighten two wires together. When I come out I was a first class electrician, and while I was working there I applied for a Marine wireman's license which I was granted and I carried that up to four years ago, and when I was seventy-five I quit renewing it.

Seamless tube inventory

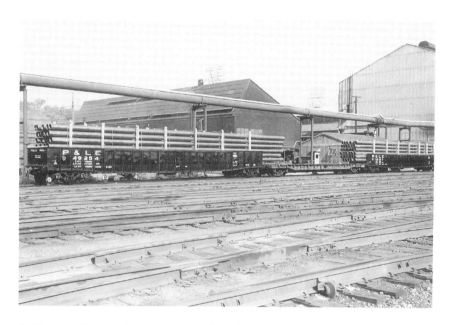

Railcar with seamless tubes for shipping

Former seamless tube site

ALQUIPPA WORKS
SEAMLESS TUBE MILLS

Production Data

Annual Capacities and Operating Statistics

Unit	Size Range of Product	Annual Capacity
30" round Mill	3-1/2" to 10" Rounds	600,000 N.T. Rounds
14" Hot Seamless Mill	5-1/2" to 14-3/8",	
	22' to 46'	300,000 N.T. Finished Pipe
6" Hot Seamless Mill	2-3/8" to 6",	
	22' to 44'	180,000 N.T. Finished Pipe

Equipment Data

Unit	No. Of Units	Description	No. & Size Of Motor	Speed-RPM
30"				
Round Mill	1	2 Hi-Reversing 3 Stands	1-3000 HP	65/130
	3	Pickling Vats-Sulphuric Acid		
	1	Gouging Machine		
	2	Billeteers		
	3	Furnaces-Coke Oven Gas & Fuel Oil		
	2	Hot Save		1450
14: Seamless	1	2 Piercers-38" roll diameter	1-3000 HP	80.8
		1 Plug Mill-38" roll diameter	1-1500 HP	67.5
		2 Reelers-30-1/8" roll diameter	1-500 HP	106.4
		1 Sizing Mill 5 Stds. 28"roll diamter	1-400 HP	33.0
		3 Heating Furnaces (Billets)		
		1 Abramson Straightener (5-1/8"-14-3/8")		
		5 Pair Cut-off Machines		
		8 Pair Threading Machines		
		2 Hydrostatic Testers 1-10,000 PST, 1-3,000 PSI		

(contd.)

SEAMLESS TUBE MILLS (Contd.)

Unit	Units	No. Of Description	Of Motor	No. & Size Speed-RPM
6" Seamless	1	1 Piercer-38" roll diameter	1-2000	77.31
		1 Plug Mill-28" roll diameter	1-1000 HP	98.00
		2 Reelers-30" roll diameter	1-350 HP	150.70
		1 Sizing Mill-5 Stds.-18" roll diameter	1-250HP	44.57
		1 Heating Furnace (Billets)		
		1 Reducing Mills-12 Std. each	1-400 HP	45.0
		1 Stretch Reducing Mill-16 Stands	200 HP	1600.0 Max.
		1 Hemphill Straightener		
		1 Sutton Straightener		
	3	Fiar Cut-off machines		
	4	Fair Threading Machines (3 Stamets - 1 Cridean - Single End)		
	2	Hydrostatic Testers 1-10,000 PSI, 1-3,000 PSI		
	1	Normalizing Furnance		
	2	Upsetters		
Seamless Annex Specialities Dept.		Size Range 2-3/8" - 10-3/4"		
	2	Furnaces-Walking Beam- Coke Oven or Nat. Gas		
	1	Quench Unit-760 Nozzles-100/125 PSI		
	1	3-Stand Sizing Mill		
	1	Sutton Rotary Straightener		
	1	10" Upsetter		
	5	Turret Lathes		
	1	Thread Miller		
	1	Thread Electro Processing		
	1	Hydrostatic Tester-20,000 PSI		

Reference 2

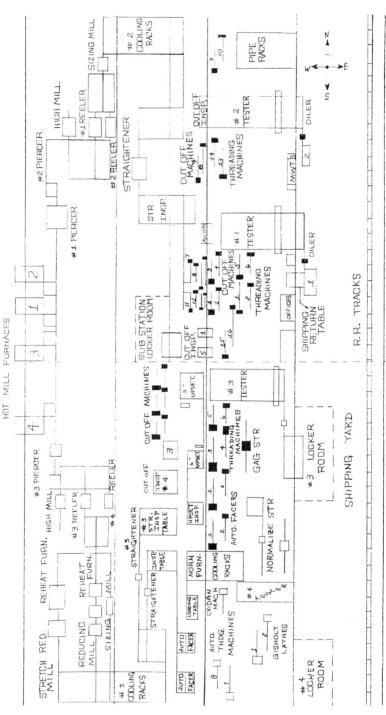

A-1-14" SEAMLESS HOT MILL

A-2-6" SEAMLESS HOT MILL

Reference 2

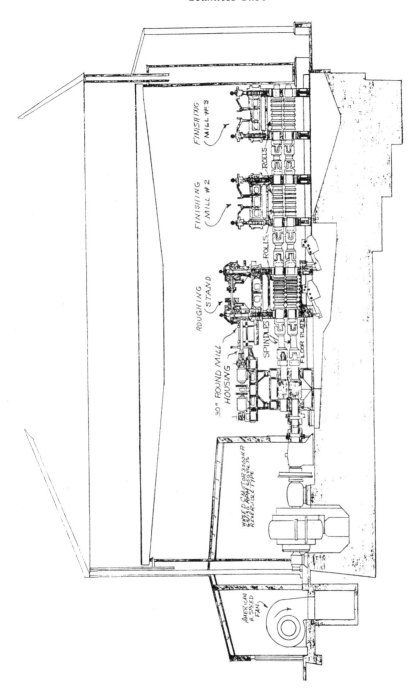

CROSS SECTION OF ℄ THROUGH 30" ROUND MILL

FINISHING MILL #3

FINISHING MILL #2

ROLLS

ROUGHING STAND

ROLLS

30" ROUND MILL HOUSING

SPINDLES

FLOOR PLATE

WESTED CORDITOR 3300 H.P.
60/30 RPM 650 VOLTS
REVERSIBLE TYPE

AMERICAN
HI SPEED
FAN

Reference 2

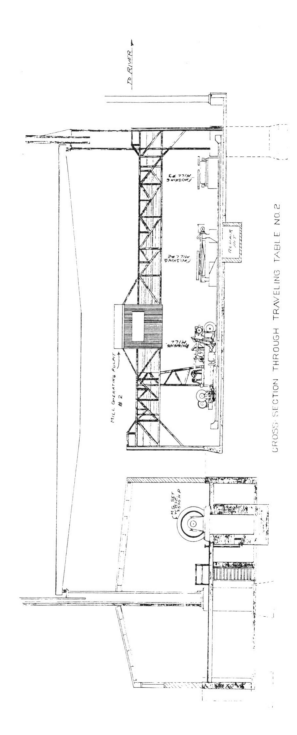

CROSS SECTION THROUGH TRAVELING TABLE NO. 2

Chapter 11

ℰℐℂℛ

Tin Mill

H is appearance was one of a younger man — spry, self-assured with an optimistic rendition of his life and times — from skipping school to the busy Tin Shop floor.

My name is George Ferezan. I'm sixty-seven years old and was born in Monaca. My parents were Romanian. I worked for forty-six years down there. The year I was hired was 1944, and at that time, I was making ninety-eight cents an hour. I retired in 1990, but can't remember fifty years ago. I can't remember how much we were making before retirement, I was making a lot — twelve, thirteen, fourteen bucks an hour, but can't remember. I worked in the Tin Mill department.

Well, in the Tin Mill, we got all our coil, raw materials, from down the North Mill. They used to roll their own steel, J&L was an integrated steel mill. We made everything, rolled our own steel, got our own coils from the Strip Mill, and when they brought it to the Tin Mill they rolled it. They cleaned it, rolled it in the Tandem Mill, tempered it, and then they stuck it through the tin coating process.

When I first came down there they didn't have any automated tin lines so they used hand machines and that's how they made their tin, with old manual tin machines. So we ran steel through the line and

George Ferezan

coated it. It used to go down through a tub of tin, molten tin, and it used to come up in through this machine. They're called brushes and it use to wipe it, put a coating on the steel, and that's how you made tin plates. The tin plate didn't rust.

And they used to run it through what they called a branner, and that would put a polish on it, shine it, and it would come out the other end. This was all slit, already. It wasn't in coil form. After they built the tin lines around 1940, '41, '42, they had tin lines in those days, but it wasn't that good. But then I went away to the service and when I came back, well, the tinning process was there. They had two good lines running and it's good stuff. They made good stuff down there. We made the best tin in the world, we had the fastest Tandem Mill in the world right here in Aliquippa. We just had a good work force.

Well, you know what, I think LTV came in and bought J&L out. I still think them guys sold us out. J&L was a good company, J&L was a good company to work for, but when LTV came in . . . This guy that bought it, Ling, his name was Ling, when he bought that, he even said that he didn't no nothing about running a tin mill or running the steel mills, and they just run it to get what they got out of it and that's it. They shut it down and they tore it down. That's what gets me, they tore it all down.

They only left a few buildings standing, that's what really hurts, because you go across the river and the World Processing is putting in this little mini-mill in over there, and they're making tin plate. They rolled steel and they want to put in the Strandcaster, and they want to roll their own stuff and all that jazz, and we had all that here at J&L. They knocked it all down and rebuilt it across the river? I think over there them guys are part union and part not. LTV was going in with a Japanese company and they're going to build a mill down in Kentucky, and it's supposed to be all non-union.

I don't know the reason behind knocking down one plant and putting up a new one. Nobody knows. They got everything they wanted out of this mill, they ran it to the ground, they had everything down there. Politics has a lot to do with it, you know, the environmentalists kicked their butts down here. That had a lot to do with it.

We lost our Byproduct Department on account of that. The people across the river used to holler all the time because they would see all that black smoke coming over the mill. People in Aliquippa put up with it because it was our mill, you know, then the people across the river always got mad cause they had to sweep their porches everyday. You know as long as they saw smoke coming out, everybody was working. Now, nobody is working. No smoke, no work. They got about 600 to 700 guys in the Tin Mill, and maybe around 400 or 500 guys in the 14 Inch Mill and that's it, down from 15,000 at one time.

They get their steel from different mills now. They bring it in and just finish it there and that's cheaper. We get a lot of the rolled stuff comes in from Cleveland, and we get some from Indiana Harbor 'cause LTV has a big tin mill over there. They built a big tin mill over there, while we were still here. The tin mill has always been a money maker for J&L and LTV. We always made the best, we made more different grades of tin plate then any mill in the country. More, and you could get as much as you wanted, you could get a box — what they called a box — or you could get tons and tons. They would sell you anything you wanted in them days.

When I first came into the tin mill, I ran a branner, what they call a branner. That used to clean and polish the plates and that was my first job in there. I worked four to midnight, I'll never forget. In fact, the first day I ran my gloves through there, through the machine. I tried to catch it and the boss saw me and he gave me all kind of heck. He gave me all kind of hell, he said, what's the matter, you trying to lose an arm or something? I said, my gloves got caught. He said, buddy your gloves aren't your fingers. That was a good, quick lesson for me early on, and I never stuck my hand in there again.

From there I went to . . . I did odd laboring jobs in the Tin Mill, and then I went up into the tin house and that's where I was a skid boy, they called us skid boys. These machines I was telling you about, when they made the tin and used to come out the other side and would come in piles. We used to pull these out, when the piles were high enough, and the guys used to come along with a tractor and pick them up and take them. We used to set this up again and roll back in the machine so it would just start piling up again. That was my first job in the tin house.

Then after that I start driving a tractor. One afternoon they asked me if I would like to drive a tractor, and I said, sure. He said, I'll give you a guy to work overtime with you, a couple three hours, so he can show you what to do, and that was all. That was my orientation, and that's how I started driving tractors.

The tractor used to move tin around. I worked downstairs, we had an upstairs and a downstairs tin house, and I worked in the downstairs tin house. That was number one, and I used to go down to the pickler and pick up these big tanks of acid and they had raw steel in them, you know, the tin. We used tanks and acid that used to clean the tin, and then I would take and put it in these tanks and carry it with these tractors. They use to call them boschs and they use to take them up to the machine and slide them alongside the machines. So these guys who fed these machines, they used come down there with a pair of tongs and took out about yea amount of steel and stick 'em in these machines, and they go down in these tin machines and come up to these brushes that brushed this tin. You have seen the tin! That's how they made tin plate in them days.

Of course, before my time tin making was a heck of a lot different. It was all by hand. That's what they called Old Hot Mill, Tin Mill — Hot Mill, that was tough. They use to do all this by hand. All this stuff they use to do by hand, even the big units that used to come on the . . . you know, the steel they made the tin plate, you used to have to crank this up there, practically, manhandled it to get away from the machine. That was my first job. My second job, you might say — and I did that for about seven or eight years, of course, I went to the service for three years and that all counted toward my seniority and when I came back I got the same job — I was a tractor driver.

Good day? I had a lot of good days. I liked my job, but never liked my job after LTV got there. When did they buy the place? Probably around 1987 or '86, somewhere around there. I never liked the mill after that 'cause you never knew when you were going to have a job. Wanna talk about stress, that was stress.

You know the worst day I'd ever had down, there I was working on that branner and they used to have a turn table, you know, you fill-up one side then you turn it around to the other side , you start shoveling your plate through there, and you didn't have to wait. The tractor would come up and get that one side that was full. You always had one empty side. I don't know, they used to have sides that used to come off, you know, where the tractor used to come up, and I remember that like it was yesterday. I took the side off and I took the other side off, and the load shifted. It came right down on me and there I was, sitting with the load on my knees and I couldn't move. The weight was about 800 pounds, but that wasn't the worst part. The worst part is that stuff is sharp and can cut you in half, you know, and there I was, couldn't move, couldn't do anything.

A guy came around — he was a millwright from upstairs — he come down and I hollered at him and he came over and got a bar and lifted from underneath that thing. And when I did that plate just went skipping across the floor. If that would've hit me it would've killed me, it would've cut me in half. That's the worse day I ever had. I almost got killed.

I do remember a guy getting killed as a pipefitter. He got killed at the Roll Shop. I'll never forget that day. I was running a crane and the next thing I know this guy, his name was John Rockavich, I'll never forget that a pipefitter was just coming on a turn and nobody . . . You're suppose to get clearance, you know, for these guys. When they go work on something there's supposed to be clearance, when you are around cranes or something, and nobody seemed to do this. This guy was up there working on some pipes and this guy in the Roll Shop, John, came down with the crane and he just got this guy between the crane and the wall and it pulled him and down he went. He hit the floor and broke his neck. He was probably dead before he hit the floor.

I can remember that like it was yesterday, when John came down out of that crane I was sitting up in mine, you know, watching. There were fifty people around and they walked him down, you know, questioned him. I was watching all of this from up in the crane and I felt so sorry for that guy, it was just like they were crucifying him.

A lot of nice guys I worked with . . . I, that's the only thing I miss down there, the camaraderie. As far as the mill goes, I don't miss that. A lot of us socialized together. I still have a couple of good buddies that I hang around with. I made some good friends at the time, we had a lot of good times down there.

You know what, there are a million stories that come to mind. You know, my mind is a blank, it would probably be too dirty.

You know, when I first went to the mill I was sixteen years old. I had been playing a lot of hooky and my dad finally caught us. I missed seven weeks of school. I worked in a show at that time. I was an usher, head ticket taker. Mom and Dad both worked. They worked daylight in the mill and they would go to work. I would come right back home and in the afternoon I would go to the show, and after the end of the school day I would come home and change into my white tie and shirt and go back to the show. I worked for Temple Theater on Franklin Avenue. Right there where Shiflet's is now, down the street.

I did that for seven weeks, and my brother said I had a broken leg. They were always asking, where's George, you know, he broke his leg. Back then nobody questioned it, but that finally ended and I got caught. That's it, hey, you either go to work or you go to school.

I chose work, I was sixteen on the 5th of March and on the 11th of March, 1944, I went to work three to eleven, I'll never forget it. The first day, that was when I stuck my glove through there.

That was a bad day. The foreman that I had said my brother worked with him at one time. You know, he told me then, you'll never be like your brother. I probably heard that a hundred times in my career down there. My brother was a boss, and I heard that from different guys, you know, you'll never be like your brother. That's too bad. I'm me, that's the way I operate. I was kind of independent that way.

If I could, I would've changed the stress. You know, what I don't understand these companies, you know, they put so much of a load on you and, you know, when LTV bought the mill that's all you have here. We're going to shut it down . . . we're going to shut it down. I don't understand that. Even when I first went down there they told me the tin mill was going to go. It's gone along with the forty years after that, you know, and it's still there. But that's all you heard.

You never know how lucky you are, you got to tell them kids today. They got so much, it's so easy. Kids don't care, they really don't. They got everything, they don't care. You kill somebody today, they don't care, I don't understand that, you know, when I was growing up, man, if you step out of line a little bit you were gone. You were gone. My mama told me she was going to throw me in Morganza, that used to be a kids school, a reform school.

That's the kind of threat that used to get to you, but not now. You walked the straight and narrow, you know, because you didn't want to go to Morganza. You got all kind of stuff coming out of that place, you know what I'm saying. So you tried to walk the straight and narrow.

Strand Theater

Well kept houses on Franklin Avenue

Old time tin shop

The tin line

He was a large man, with a boyish appearance, his green flame retardant completely covered with grease and fine black grit productive with the work of the day.

My name is John Carr. I am forty-seven years old and was born in Steubenville, OH. I came from Steubenville to Aliquippa when I met my wife. I was going with her, played music in a band, and then I came here.

I have twenty-two years of service and was hired in 1973, November. I was making 10,000 or 8,000 dollars, something like that at the time, and now about 40,000 dollars. Currently I am working in the Tin Mill as an electrician, hourly.

I started in Blast Furnace. I was a laborer in the Blast Furnace for a year and then went through apprenticeship, electrical apprenticeship, and graduated from that and was assigned to the Tin Mill, and I have been there since 1978.

The Tin Mill coats tin, puts tinplate on steel, and that's the basic product to get out tin products for can companies and for automobile parts and things like that. They bring raw steel — not raw, but uncoated steel — here and then we coat here.

Well, what it runs through is the Tandem Mill. A Tandem Mill is a five band line, it sizes it. It runs it from there to Coil Anneal (CA) or Batch Anneal line. If it goes to CA line it goes through the cleaners. It goes out of there to the cleaners, to number three stand or either Number Four Mill, depending on what product it is. From the Four Mill it goes to either number three or four coil preps. Basically, I think it is three coil preps that are still running.

That trims it and then they bring it through the lines. Three Tin Line or Four Tin Line, and with Three Tin Line we can, now they upgraded it. We can do a lot more with the product than we were before. But we're a finishing mill from that point on. I work in Tin Line South.

What they interpreted years ago as a motor inspector. Check on the lines, the brighteners. My job is to make sure the brightener runs, the motors run, that there is lighting around there, basically any electrical problem there is.

They're all pretty good days. I basically can't pick out no one day that was better than any other.

I had one mishap in the mill and I broke my hand. That was about my worst day. I was out working on a shearing machine at that time and got my hand caught in a roll and that was my worst. Any injury is probably the worst day of your life.

Well, North Mill to South Mill, that's what it used to be called. North Mills had all the hot areas, where the hot steel came out of. We basically don't do that no more, just mostly cold steel down here. Personally both areas are about the same to me. Working conditions are different, but people-wise it was a little different down here than it was up north. A little more friendlier there, I would say, and a little more together. Down here in the Tin Mill it was a little less friendlier when I got here. It was all right, I'm used to it, it's easy to get used to.

Yeah, working conditions are so-so. They're pretty good now compared to when they first started, but I couldn't really go back to qualify how much different they are. Working conditions now are pretty good. I have no problems with it. Yeah, right now hourly and management are in a situation. Obviously people are pulling together, we're still in business. Well, it takes both parts to make it happen. It's not a one-sided issue. If you don't have both parts working together you won't have a mill, and that's the situation now. Union and management try to work hand in hand.

A few bosses, I have different bosses. I think a lot of our general foreman that we have down in the Tin Mill. He's a person that I would say is a pretty good guy. Workers that I work with are pretty good people, they helped me a lot when I first came in their gang. A lot of them — there is a whole bunch more — but just to name a couple . . . Other areas I've been I liked. Well, one guy is down here now, Johnny Dzumba. When I worked as an apprentice down through Strip Mill, he helped me a lot, and then Floyd Thomas. I have had some good workers, I have had good rapport with the guys I work with everyday now, and my turn foreman is not too bad, he's pretty good.

Well, that to me is a little sore piece because I think we are getting older and we are getting less guys, and I think the company needs to look at that factor that we are getting older and we need more help. Their opinion is that they can do it with less people and still get the same amount of product out, and things like that. I think we should look at that point of view and get it straightens out, and that its pretty good. Well, shifts and industrial work, well, this is actually my first year to having to worry about shifts in a long time. I worked basically daylight, because I was a young person and a lot of guys didn't want that job that didn't make no money, so I've now started working turns because of the need. They don't have the people, so now I do shift to shift. Getting used to it, I didn't mind midnight, which now I don't work because we've taken a guy off the unit here. But I didn't mind midnight, but I hate four to midnight, three to midnight, whatever you want to call it. I hate the afternoon shift but the other two turns don't bother me at all.

It seems like I don't do nothing but get up and go to work, come home and go to bed. I'm just in that position that I don't get much done. A lot of people can do that turn and it benefits them, but me it just seems that I am out of it when I'm on that turn.

The job is different now. What we did when I was doing by hand is being done with computers today, and other things, and making differences that way. The thing that I could tell anybody young is to get the best education you can. Try to find something you enjoy, and it'll make it easier for you to do in the long run. A job is what you enjoy doing, sometimes, you have to think. It takes a little less money to enjoy what you like, but you'll last longer at it.

About the mill? I hope that the future last a little longer. I'd like to see it last another twenty or thirty years. We do get some people with less experience to get a chance at this, and I hope that the union and the company can become a working group together that helps both the ends meet the means, which means to me where each person can make an equal amount of money and be satisfied with what they are doing.

Finished tin rolls

The tinning line and inventory

Current Tin Mill

ALQUIPPA WORKS
TIN MILL

Production Data

Annual Capacity - Packed Plate
(Estimated with Continous Annealing line and New
#4 High Speed Halogen Tin Line in Operation)

Electrolytic Products	540,000 N.T.
Hot Dipped Products	30,000 N.T.
Black Products	<u>30,000 N.T.</u>
Total	600,000 N.T.

Reference 2

Equipment Data

TIN MILL (Contd.)

Rolling Mills

Mill	No. Description	Units	Roll Size Work	Back-up	Make	No. and Size OP Motors-HP	Speed Max FFM
Cold Red.	42" 5 Stand-4 Bi	1	2½"x42"	53"x42"	Mesta	(1-1750 HP 1st Stand (2-1750 HP 2nd Stand (2-1750 HP 3rd Stand (2-1750 HP 4th Stand (4-1000 HP 5th Stand (2-300 HP Reel	6250
Temper	42" 1 Stand-4 Bi	2	19"x42"	39½"x42"	Mesta	(1-250 HP Tension Reel (1-400 HP Mill Rolls (1-400 HP Delivery Puller	3600
Temper	42" 2 Stand-4 Bi	1	19"x48"	49"x48"	Bliss	(1-300 HP Entry Puller (2-400 HP 1st Stand (4-300 HP 2nd Stand (4-300 HP Delivery Puller (2-300 HP Tension Reel	6000

TIN MILL (Contd.)

Other Equipment

Unit	Description	No. & Make	Speed
Continous Pickler	4 Pickle Tanks	1 Meats	2000FFM Entry 600 FFM Delivery
Electro Cleaners		1 Wean 2 Meaker	2200 FFM 1800 FPM
Continuous Annealing		(See Description Below)	
Coil Annealing		13 Swindell-Dressler 5 Lee Wilson	
Coil Preparation		2 Aetna Std.-1 Wean	2200 FFM
Coil Rewinding		1 J&L	600 FFM
Electro Tinning		(See Description Below)	
Shearing		1 Mesta 3 Hallden 1 Hallden	600 FFM 650 FFM 1000 FFM
Hot Dip Tinning	5-64"Machines 5-75"Machines		

TIN MILL (Contd.)

Continuous Annealing Line
Started Operation on March 18, 1960

Type of Line	Radiant Tube, Dual Gas Fired (Electrical Resistance, By-Product and Natural Gas)
Line Delivery Speed – Maximum	2,000 FFM
Number and Size of HNX Machines	2-20,000 Cu.Ft./Ht. Each
Computers	General Electric

Electro Tinning Lines

	#1 Line	#3 Line	#4 Line*
Type of Process	Halogen	Halogen	Halogen
Number of Plating Cells	6	24	12
Plating Power Amperes	36,000	180,000	120,000
Rated Line Speed: (25"Plate) Max.	900	1800	1800
Width of Product	35	37	37

*No. 4 Tin Line is expected to start operations January, 1961.

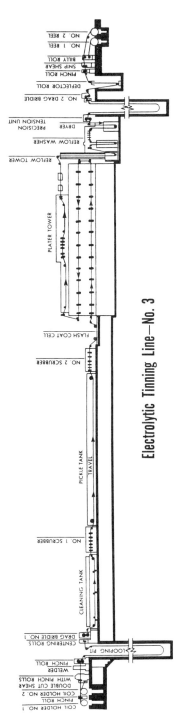

Electrolytic Tinning Line—No. 3

Chapter 12

ℰᑑᑕℛ

Rod and Wire

H e proudly held his compilation of the 1959 Labor Day picnic —
leather bound, dusty, black and white photographs. It was a
microcosm of America — like a Life magazine edition with hotdogs,
pony rides, raffles and the American dream.

My name is Michael Rebich. I am seventy-nine years of age and was
born in Aliquippa, PA at Rochester Hospital. My nationality is Serbian.
I worked in the mill for thirty-seven years and retired when I was sixty-
three. Well, I got hired in the 1930's, right after The Depression.

I got more deductions now than I made the whole year when I started,
not very much. I retired, when I was sixty-three. That was, I think '70,
something like that, 1970. I was making around 1,500 dollars a month
salary at that time.

I started out in the Wire Mill department at J&L Aliquippa. They
made your raw wire, you made your raw wire from rods. I worked in
the cleaning house, where rods would come from the rod mill. Then we
would clean them and put lime on them and then the wire drawers would
draw them down to whatever size they needed. They sold wire to
everybody you can think of. The marketable products were nuts, bolts,
and welding rods, but mostly it was wire. What eventually happened to
the wire we don't know. It was used for a million things there.

Michael Rebich

My first job down there was an acid-tester. In the cleaning house, where they cleaned these rods, they used sulfuric acid. We had to test the intensity of the acid and add or subtract, mostly add on, and that was it mostly.

Now, before I went to the Rod & Wire, I was a general laborer. I worked every place. Manual labor, everyplace you can think of. Then I went to the wire mill, and from there, worked, even drew wire. I made dyes, worked in the dye room, and made the dyes. The dye actually would be put with the wire to add the screw threads. It's smaller, round, like a half a dollar. You'd have a dye like a half a dollar, and it had tungsten carbide in it, and then you would have to shine 'um and polish 'um up. Put it to size to whatever size it needed. In the wire mill, when I worked in the dye room near the wire mill, wire drivers would come up with a certain size, and then you'd have to give him the dyes that fit. Come down to that size because you couldn't come down to . . . With one dye to one size, it gradually comes down, so we use to make those dyes in different sizes.

Well, when I went into management, I was clerical. Taking care of spare parts, inventory, and whatever I had to do — type out orders. The boss would give me the orders and then I gradually took the orders

myself. My immediate boss retired and I worked into that job. Got into management and then I stayed in the Rod & Wire department. Because it took care of all the rod and wire supplies and products.

Your foreman would come in with orders — they want this, they want that, and you'd have to write them up and send them out and wait till your boss signs them. Sometimes it happened they wouldn't sign the orders, and then when you had a breakdown, you got your rear-end chewed out because you didn't have the part.

Younger Michael Rebich

When J&L was there, you could talk to your boss. I was with the Quarterback Club and my boss was with the Quarterbacks, so I used to tell him, Bill, you never signed that order, and he would accept that. But when LTV took over they wouldn't accept it.

Oh, it changed drastically. They wouldn't buy anything. In the Wire Mill they were cannibalizing the Wire Mill machines. When they needed a part they would take it off another machine. Machines, they're sitting there, they're no good because the parts are gone. Well, when LTV took over, they started to do that. Before LTV took over they were little by little not buying parts, and then when LTV came up they wouldn't buy any parts, and we used to do a lot of repair work. Our people would

Examining wire spool

do all the repair work here. Before LTV came in they weren't allowed to ship any work orders out. Certain ones they were allowed to ship, but when LTV come over, they would have you ship everything out for repair.

They contracted out most of their repairs. I guess, for money saving. No, because I'm sure the people we had were able to do this kind of work. I know I used to see a lot of work out of here wire drawing blocks. Big, scale them down, and harden them, and resend them all out and then, well, that's one reason I quit.

When LTV came in, I had a foreman, and the foreman usually comes in and gives you a work order that they want and most of them were . . . I would say very few were knowledgeable. There was one guy here and he was Serbian, he knew everything. He would come in, give you these orders, and that was no problem if you knew what you wanted. But then he retired early, surprisingly. I thought he would stay there for a lifetime, but then these other foreman come in, they didn't know what they wanted, but this other guy was knowledgeable. They used to go to him and he would tell them, but when Nick was gone they come in and tell me they want this, they want that, and I say, what is it? They don't know.

It would make the job harder, and then the last time, what made me retire, we had a breakdown. This new boss, didn't let us have the parts, and I got my old boss, when it was J&L he use to be able, use to be able to talk to him, but when LTV took over, we couldn't talk to the bosses, I guess pressure was them too. He wouldn't accept no excuses, so one day I'm in the office. I came in early and my boss comes in, his name was Don. I said, Don, I'm retiring. He said, it's not anything I did? I said, no, just tired of working here, I quit. That was when I was sixty-three. That was one of the reasons.

A good day in the mill was a day when nobody chewed your ass out. I don't know, you had, well, you had a lot of good days, you know, they weren't all bad. But to pick one out as the best day, I wouldn't know. Maybe the best day was when they gave me the job.

First day on the job, well, when I got hired into the mill? Very first day, when the Open Hearth broke down and I was in labor. I was working — going to school and working midnight my last year — and the Open Hearth broke down. Those were dirty jobs, we worked in all the rat holes down there and it was hot, dirty, and so forth, and those . . . like I said I worked in every hole up there.

But then they had a softball team there, and I was a pitcher, and I think I was pretty good. Well, that's why they gave me a better job. Then they put me on a Red Cross team we just go around putting in time, you know. How I got the job, one of the union men, Sam Urick, he was a union representative and he knew me.

General labor use to go out there, you remember, the docks. They picked out people to work. Each foreman would need somebody so they say go out and pick out five men to work. So the foreman would go out and pick five men, you take five, three, or four, nothing was left. I'm on my way home and Sam said, hey, that kid can pitch, come on back. Then I worked everyday.

Then I worked every day, that was when J&L had teams down there. In fact we won, one year we won the championship. Not only did we win the championship, whenever we played ball they use to let us go out of the mill then come back in, go out and play ball and come back in. Otherwise we couldn't have done that so this is under J&L now.

J&L, I thought they were a good company to work for. They had their faults but overall it was a good job. Yeah, they were better, they were definitely better than LTV, let's put it that way.

The worst day's when the Open Hearth went down. Then, you had your dirty jobs. You had your hot jobs, you worked in the slag, you had slag-pockets in your Open Hearth. Slag, you had to break those up, you had your what they called the checkers. They were wire bricks across this way where air would come out and those were hot. You had to wear wooden shoes, wooden shoes to prevent your feet from burning, and you got dirty, too.

No, all I remember would be again coming back to those Labor Day celebrations, that was when I was with the union. We had a lot of opposition, when we first started the Labor Day celebrations. Well, you know you have some people who are opposed, some people are opposed to everything. So I had a horse show coming in. You might've saw it in one of those clippings — that so and so horse. 10,000 dollar horse show. Well, we ended up spending 50,000 dollars more. The more we make, the more we spend out. We had opposition then, but then afterwards they, you know, after that they let us spend as much as we wanted decently. This was union money.

This Labor Day celebration cost 50,000 dollars, yeah, at the end. I would say on our fourth one because there were a four or five day celebrations. Oh yeah, the workers liked it because, you know, Labor Day, we stressed, keep the people in town, keep them off the roads, and this was one of the things. Then we had old time prices — nickel pop, ten cent beer, five cent, those prices, and we still made money. We even raffled off a car, a nickel a ticket. We used to sell them a dollar a book and we made money off of it surprisingly. Yeah, a 1960 Valiant automobile, we made money on that, a nickel a ticket. We didn't make too much, but we made it. That was back in those days.

Well, no regrets, I would say. I had opposition there when I picked committees and handled a lot of things before I took over. I knew who was good for it and who wasn't. Tony Vladovich just got out and Nick Mamula took over, see, so I had a lot of Vladovich's people and Joe was one them. Then I put them on committees, and right away I had opposition from Mamula's people, and then one thing I'll give Nick credit for: I told him you want me to a job and you want me to do it right, these are the people I got that you can depend on.

I can give Nick credit for that. He said, go ahead, Mitch, take who you want and I had a lot, that's where I had a little opposition. I had a lot of Vladovich's men and, you know, whenever a guy goes out the men change. Then whenever Mamula went out I ran into the opposition. They wanted me to handle Labor Day, because Mamula was out, so I told them I would handle it if I'm allowed to hire my own people.

They wouldn't, so I quit. Now you see the difference, Moe Brummitt was the president. He took over and his people wouldn't let us do that, but again I give Nick credit. He allowed me to do that cross party lines, but this other guy said no, so I quit.

Well, Nick Manolovich, I remember he was a foreman, mechanical foreman, and knew all the parts in the mill. He made the job easy for me and Don Hatfield, my foreman. I give him credit for starting me off on playing baseball. When I retired, my boss gave me a dozen balls. They were golf balls they painted them orange so I could see them. My boss, you know, whenever it was salary issue, whenever, if I worked overtime, I got paid, not like I was exempt. Yeah, whenever the union got a raise, I got a raise. Always a little bit more, and the boss would come up. He would call me on the PA system, wants to talk to me, and that's about time I knew union had a contract. The way he would do it he would throw you a slip of paper. You opened it up and this was what I was going to make for the next year. So I always told him, I said, Bill, is this my automatic raise or is this something extra. He said he don't have no money, he's got to take care of some other people. It was always that.

He always give me automatic raise because I got five, ten percent more then he did. I don't remember, so that's what he would say. So when I retired, Bill was my boss, he gave a little speech. So I told him I thank my boss, my immediate boss. I thank them all for helping me out, and then I thank my boss for keeping me on a "low-income basis" during my retirement.

And my boss said how come you have guts enough to say that. I said, look I'm leaving. And that's the truth, I would always ask him when I got that slip of paper, with my retirement package. How about some extra. He said, we don't have no money, I got to take care of this and that. So that was a little amusing there.

Well, well, when a guy got hurt, when a guy got hurt he had to go to the hospital. It happened, that happened to me. No, it was human error, or . . . Well, down there injury, the responsibility rests with the injured employee. That was their favorite phrase. So I was working, my daughter got sand in her eye, so I had to work night turn. I worked night turn and was laboring and they had coils of wire on buggies like this and peter tractor would come over and pick them up. So they had two bundles on there. I got a hold of the tractor and he came in. He didn't have to come against the truck but he went and got stuck and I pulled my hand out and that was it. So while you was there you need to wake up, and stay alert.

There was another wire drawer, he lost a couple of fingers, he got caught in the wire. So you had to be careful.

Well, when I went on salary, it was steady daylight — Monday through Friday, but until then I didn't like it. I like daylight 'cause I was able to play ball.

No, the hardest part, one of the hardest decisions I made was whenever they called me down the employment office to sign my rights away. I'm not a union member because I signed my rights away. I had to think twice there, but then said, I don't have any further to go in labor. If I stay where I am, they would put somebody up above me, so that was a hard decision to make. Because my boss even afterwards said, Mitch, you're still a union man because I took care of them as much as I could. Millwrights, you know, millwrights used to work under us a lot.

There's a hundred things you could think of — relations, the trust between union men and companies. Well, it seemed like the company would do something and then they would look for a reason they were doing it. Well, what are they doing, do they got an ulterior motive, if they give you something? They want something from you. I would think a lot of them did, but that runs throughout unions and company relations, but again we come back to J&L.

We used to have golf, we used to have bowling leagues on J&L time, and used to have a banquet at the end of the season. You would ask them for money and they would give you so much per each participant and at the banquet there. But then after J&L, well, it was even before LTV took over when they cut that stuff out. They wouldn't give, but up till that point, Labor Day celebration, I need electricians. I needed carpenters, and they all came out of the mill, they even furnished lumber and everything. It all came out of the mill. Oh yeah, that made the men happy, but again I say J&L per-se was a good company to work for until things started going bad. Why, I wouldn't know, they shut the Nail Mill down because they were importing the nails. Then they shut the Rod Mill down because . . . I don't' know why.

Last day in the mill, I told you about that. I got my rear-end chewed out, That's it, I quit. I said who needs this? Again, that Nick Manolovich, that foreman knew, he quit. He knew more than some of the big bosses. The hardest part was getting your rear-end chewed out by management, when they didn't know what the hell they were doing, they didn't know their ass from a hole in the ground.

Get an education and stay out of the mill, but the mill, if you get in, it pays more than they do on the outside. You don't have minimum wages, although the wages they do have now, they're minimum compared to what we were making.

Sure they were, they used to make, you know, when the seamless was going strong, they use to work double shifts, six days a week. And they were making . . . and then this come in the paper where they say steelmakers, a steel worker was making 50,000 dollars a year. But they didn't say how much time he had to put in to make that 50,000 dollars. All they said was 50,000 dollars. Some of them made 50,000 dollars, but they didn't say how many hours he worked, how much time he put in, time-and-half, doubles and all that stuff.

Wire product shipment

Wirespools

Wire production line

The end of the Wire Mill

Wire Mill demolition

First Manual Labor Day celebration (1959)

Marching band

ALIQUIPPA WORKS
ROD & WIRE DEPARTMENT

<u>**Production Data**</u>

<u>Annual Capacities and Product Sizes</u>

Unit	Size Range of Product	Billet Used	Annual Capacity
11" Rod Mill	600# Cls. .218"-.656"	2½" Square	362,000 N.T.
Wire Mill	Wire, Fence & Nails		315,000 N.T.

<u>**Equipment Data**</u>

Unit	No.	Units	Description	Drive	Speed
11" Rod Mill	1		United 2-Hi Continuous	Electric	5000 FPM (Ave.)
		8	Roughing Stands	4- 500 HP	
		8	Intermediate Stands	4- 800 HP	
			(Includes 3-180°Loops)	1-2000 HP	
		6	Finishing Stands		
		6	Laying Reel Coilers		
		6	Pouring Reel Coilers		
Wire Mill - Cleaning	1		Straight Line Rod Cleaning Unit – 5 steel acid tubs		
	1		Cleaning House – 6 wooden acid tubs		
	1		Fine Wire Cleaning Line – 3 wooden acid tubs		

Reference 2

Equipment Data (Contd.)

Unit	No.	Description	Size Produced	Spindle or Block (Finish) Speed
Wire Drawing	4	18-Block Morgan Benches	.072"-.225"	43 to 60 RPM
	4	Vaughn Double Deck Motoblocs	.135"-.375"	60 to 120 RPM
	6	Type "A" Morgan Conner Cont.	.072"-.135"	120 RPM
	9	Vaughn Motoblocs	.170"-.968"	15 to 68 RPM
	8	3 Block Vaughn Continuous	.099"-.162")	
	25	4 Block Vaughn Continuous	.041"-.235")	
	14	5 Block Vaughn Continuous	.056"-.235")	
	21	6 Block Vaughn Continuous	.024"-.120")	Up to 1800 FPM
	2	Intermediate Coarse Vaughn Wet Wire Continuous	.016"-.041")	Up to 2000 FPM
	17	Fine Vaughn Wet Wire Cont.	.005"-.020")	
	2	5 Block Aetna Continuous	.028"-.062"	230 to 930 FPM
	6	8 Block Aetna Continuous	.011"-.033"	650 to 1500 FPM
Heat Treating & Galvanizing	5	Lee Wilson Annealing Furnaces – 14 Bases		
	5	Frames Open Flame Patent Annealing 30 to 36 Strands per Frame	.028"-.135"	
	1	Electric Patent Annealing Frame	.035"-.312"	
	2	Oil Tempering	.047"-.300"	
	3	Wire Galvanizing Frames		

Reference 2

Equipment Data (Contd.)

Unit	No.	Description
Fence	18	Barbed Wire Machines – 2 or 4 point
	7	Field Fence Machines

Standard – 22"-55" in height
Poultry and Rabbit – 48"-58" in height
Chick Tight – 48"-72" in height

	Standard	Galvanized
Length –	½" – 12"	½" – 12"
Diameter –	.035"-.375"	.072"-.375"

Unit	No.	Description
Nail Mill	148	Nail Machines
	4	Staple Machines
Miscellaneous	1	Bluing Furnace
	1	Nail Galvanizing Drum
	10	Straightening & Cutting Machines

Reference 2

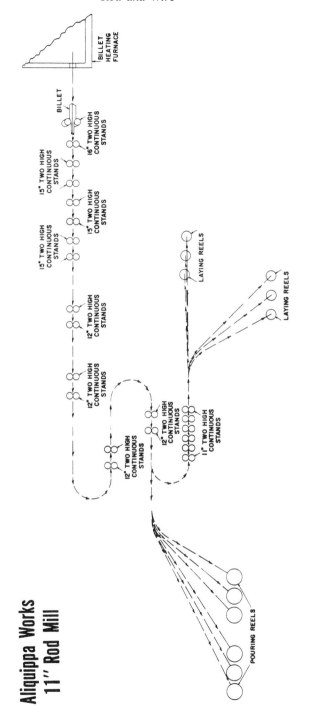

Aliquippa Works
11" Rod Mill

Reference 2

Section IV

General Maintenance

Chapter 13

ℰꙄჀᏫ

Boiler Shop/Pipefitters

He seemed almost professional in appearance — brush, cut, hair and mustache with upswept eyebrows. A man of numbers and specifications.

My name is pronounced Zelenak. I'm sixty-nine years old and was born in Aliquippa. My nationality is Czechoslovakian. I worked in the mill for thirty years, actually if they would've gave me my army time, seven years, it would be thirty-seven years. I was in World War II and the Korean War. I was hired in 1942. They took me right out of high school, sixteen years old. I went right down there and got a job. I started out at seventy-nine cents a hour and retired on July 27, 1983. I remember that day. I was making . . . we were called number one craftsman, in J&L, boilermakers were seventeen points. I was getting about fourteen an hour.

Yes, boilermakers, ironworkers, in J&L Aliquippa Works. We did riveting, blacksmithing, all that stuff as a boilermaker. Sometimes we go in the boiler and change the superheater tube. We had to crawl in there like a snake, burn the old tube out. Did you come down past J&L? See that building still standing?

Martin Zeleneck

The new building, you know, where the overhead bridge used to be in Ambridge. You come down a little further and you see that building standing. We put that building up, us and the riggers. I drilled rivets on that building. We erected the buildings, in the shop, we made everything. We could make that coffee machine standing over there. We had big machines that could bend, roll, and shape steel into just about anything. We could make a car bumper if we had to.

A boilermaker dealt with heating and cooling pipes, and that was our primary job. We changed tubes in the boiler. This was sort of steam for power production. They made electricity for J&L. Say, if number 58 boiler was down, the biggest boiler, and we had to try to get it on-line as quick as possible so J&L did not have to pay Duquesne Light Company 100,000 dollars a day.

We had our own power plant. They heated some boilers with coal, and some with oil. Pulverized coal, this coal used to come in there and they had to pulverize it. Make it like dust and shoot it through the cyclone. That was like dust, that's what the pulverizer did. They would crush that coal. It went through the cyclone and the cyclone blew it into the boiler just like that. The 58 boiler — 57, 56 they were the three big ones. They got that big number 60 boiler, they heated that boiler with blast furnace gas, too. That gas comes through a vortex to purify it, get all the dust out of it. They used blast furnace gas, they even used gas for the byproducts and the coal, and you worked in the byproducts changing the cooler tube.

So basically the boiler would heat up with coke or furnace gas, turn to steam, turn turbines, and you had electricity. We made electricity for all of J&L, and if those boilers went down — even one of them went down — then they had to buy electricity from Duquesne Light Company at 100,000 dollars a day. Number 60 was the biggest one and when that one was off line, that's what one of the foreman told me that it cost them 100,000 dollars for twenty-four hours. My position when I started, or job title, was apprentice.

For three and a half years and I had to go to school twice a week, too. We took English, math, drawing, I had to know how to read drawings. We built a lot of stuff off blue-prints and then, when I graduated apprentice school, I had to make a school drawing to scale.

You know how you hear a lot of people say, well, the Japanese and Germans and they subsidized their workers wages, but we were subsidizing also. J&L wasn't paying me, the government was paying my wages when I was an apprentice. So we were subsidizing our workers. After I was an apprentice, then I went to C-rate, then we were C-rate for, I forget, maybe a year. Then you went to B-rate and that was another year, then you were A-rate, top pay rate. We were not working independently at that point in time. We always went out in three or four men crews. In the riveting gang we had — it was four men — you had a heater, a catcher, and then if we were double-gunning, a bucker with the dolly bar, and then the driver.

The jobs we would do . . . on buildings the carpenters built scaffolds on those purlins, or supports, and those purlins are two angles, they are four-by-six angles. They were about ten inches wide, from the crane runway. We had to walk out, we had to pull our own float up, then the carpenter would start building those scaffolds. We used to pull our own float up and the carpenters would start building the scaffold, and I would have to get a sixty pound hammer and walk across that purlin out to that scaffold and drive some rivets into there, and then move to the next section and then we riveted on ladles. Those big 365 pound ladles, and some of those rivets were an inch and a quarter. We used the big hammer, the big ninety pound hammer. Well you have a sixty hammer, a eighty, and a ninety pound hammer. The ninety is the big one that's to put those big inch and a quarter rivets on. You might have four inches of stock sticking out and you have a 110 pounds of air pressure coming through that. We usually have to double gun them to get the rivet in. Then we put lip rings on those ladles and the blacksmith would work the big angles.

I did blacksmith work, too, where we would roll plates with this big machine. In other words, they had stoves where super-heated air went into the blast furnace. We put plates on these stoves and we rolled these plates in our shop. These plates were ten foot long, five-by-ten inches, and you would roll them, and the thing in the big crane would bring them up to our scaffold, and we had what's called landing dogs. We put our ratchets on there — a dog. You know what a dog is? You have a piece of scratch paper?

What we call a dog, all it was, was a half inch plate and it was made like this here, and the welder would weld that thing there, and then they would weld it here, you know. They would weld it here and to the side of the stove, and they would put a wedge in there to drive the plates so the welders could have a place to burn.

We had a machine in our shop called an octopus, it had three tips on it and it followed a pattern. Then we had a guy who would make a template and that thing would follow the thing. Say you want to cut three different things, that octopus would do that. I would say the best day would be being in the guard gang. In the guard gang we made guards for gears so that people wouldn't get their hands caught in gears and that. Well, we would make these guards out of one inch plates. I'll show you how a lot of the guards, how we made them. Say you had a big gear over here and then you had another small gear over here. We would make this guard it would be out of a one-eighth of an inch plate like that, and then we would have a trap door so that you can shoot oil in there, and then we had machines that would cut that one-eighth inch plate like a jigsaw. I use to like to do that, make them guards.

The worst was when we had to change the bell in the blast furnace. The big bell was fifteen tons and they would get cracks in there and now we're 160 feet up in the air. We had a ladder in there. They're cooking iron underneath you. If you looked through them cracks you could see the iron, it looked like a volcano, and we were putting patches on that. We had fresh air masks not oxygen, plain air to breathe and we had to put a safety belt on, but if that bell would've broke there wasn't nothing to save us and we would've burned alive. We would stay in there half hour then come out and somebody else would go in.

We were putting big three-fourth inch patches. There was a little bell. When the iron ore and everything would come up on the skip car it would dump it into the little bell. The little bell would open up then the stuff would fall down into the big bell — the coke, iron, scrap iron, and everything. Then they would just drop it down into the blast furnace 160 feet below. That was the worst job. And some of the boilers, too, changing tubes in the boiler room when it was ninety-six degrees outside and that boiler house would be 130. We'd have to crawl to get that tube out and crawl inside there and burn it out at the headers. They were four inch tubes and it was so hot. People thought it was hot last Sunday it was ninety-six degrees outside inside the boiler room was 130 degrees.

A day that sticks out in my mind, we were working down the stock house that day, three to eleven, and the carpenters were putting the big scaffold up. We were going to repair that blast furnace. This was in the 1960's. Somehow the big scaffold moved and it crushed one of the carpenters. The carpenter and foreman fell down, all the way down, 160 feet down in all that slag and everything like that. We all went home that day, we just checked out. They didn't say anything, they let us go.

Another thing, Ed Ringer down here, he could tell you this. His good buddy was a carpenter down at Byproducts, worked with those tar tanks. The carpenters were going to put a new roof on that tank, and he walked out on that tank and it gave away and he fell down into the tar. They had to empty all that tar out to get his body out of there. He died instantly.

It was very toxic. We were working on the primary coolers that day. See, what they used to do, they used to put planks out there. He was young and didn't know anything, and he walked out there on that plank and it gave way, and he fell down into the tank. I think just the fumes alone killed him.

What I won't miss about going to work was the winter time. We always worked on the ore bridge in the winter time, we had to carry salt with us when we walked up those steps cause that ore bridge was 180 feet in the air. There was ice on those steps and the steps were real steep, so we had to put salt on them to get up there. We always worked on the ore bridge in the winter time, and then it's ten degrees colder up there then it was down on the ground. I think they did that because in the winter time I guess maybe production was lower, and in summer time there was much more production. So we did that maintenance and repair work in the winter when production was down.

One guy I worked with, Johnny Andeko, he was a real smart boilermaker. I used to like to work with him because he was real patient, and I learned more from him then. Boy, he was some strong boilermaker. You wouldn't believe it, you just wouldn't believe it.

I'm telling you we had this guy, John Stasco, six foot, five, if he was living and a young man today, the Pittsburgh Steelers would have nobody to beat him. He once picked up two big men, picked them up by scruff of their necks and knocked them together. He would take a sixty pound riveting hammer and drive rivets with one hand. See, that's when them guys worked hard sledging. This other guy, John Dugas, he's another Czech. I don't know if you ever heard of Dr. Berkman. Berkman gave him a physical one time and when he stepped down, Berkman, he thought he was a Neanderthal man. Small hips and huge shoulders. Strong, you can't believe how strong these men were driving rivets, working hard.

My father was strong, his name was Martin, like mine. And he . . . you know when he came to America his brother was here and sent him money. When he had to cross Austria to Hamburg to get the ship, the Austrian army threw him in the army. He was an eighteen-year-old kid, became corporal in a horse drawn calvary, and that's during the Russian-Japanese War. That was in 1903 or 1902. Franz Joseph sent a whole army into Russia because they feared the Japanese were going to overrun the Russians and come down into Europe.

And my father was there. When he landed in New York, he walked from New York to Donora, Pennsylvania. People used to buy sacks of potatoes by fifty pound bags in those days, from potato sellers. He used to steal potato sacks off people's porches and wrap his feet when his shoes wore out. He walked from New York and got rides on stage coaches and stuff like that.

Like I said, he was in the mill right in the boilerhouse, and there was twelve-hour shifts in them days. They didn't have eight-hour shifts till the union come in. Then he'd get carbon monoxide gas from working in the boilerhouse. When, we worked in there we had a gas mask and a meter. If there was to much gas we had to get out of there. But my father they didn't have it. He get gassed, go out in the alley, lay down and throw-up, and go back in there and keep working. If he wouldn't, they would fire you.

There was the 1936 flood. The South Mill Boilerhouse was all flooded down and he knew how to babbitt or unite to metal surfaces kind'a like welding. Now they have bearings, and then they used to put the babbitt on and then the thing would start running, and you'd have to melt that stuff and shape it and put it in there, and that was like what they had today with bearings. His superintendent came up to the house and told him, tomorrow morning we need you, and they took him into the mill and they paid him twenty-four hours a day. Had a bed there for him, fed him I think for a whole week till he got that boiler on-line.

The union was very good, very good . . . very good. The union made it a better place to work, it made it safer for us, and if we thought a job was unsafe, they had what was called a safety man. We would get on the phone, call our union up and the safety man would make a determination. We wouldn't work, we stopped work right there until the safety man come there and talked with the company and got it straightened out, and they made it safe for us, otherwise we wouldn't work. The Occupational Health and Safety Administration (OSHA) would help, too. People talk about it, but there was nothing bad about the union.

My family didn't mind the work. Yeah, my wife is from Germany. Yeah, she is a war bride. During the Korean War when I got wounded, they sent me to Germany in 1961 or '62 and that's when I met her and married her. It did not affect her at all, having to work shifts. No, oh no, I worked mostly steady daylight, but every twelve weeks I would get midnight and there were a lot of guys that wanted to work midnight cause there was fifty cents per hour difference, shift differential and they would take my midnight if I didn't want it. So the guys worked midnight to make more money. You got fifty cents extra an hour, that is shift differential. I think for three to eleven it was twenty-five cents more an hour.

I would change the safety factor, if I could change something with the mill. There were a lot of places like where there was gas, byproducts. There for the longest time, just before I retired, then they started putting signs down there — danger, cancer or whatever. No, most of the guys that worked down there steady died already. Our money in our wallets would actually change color, it would turn green. We would go down there about every six months to work on the superheater tube, change superheater tube, or these cooler tubes. The money would change color, it would turn green — your silver coins in your pockets, they would turn green even if you had them in a little purse. The guys that worked down there steady, they didn't put their clothes in a locker. Everyone had a chain they used to hang their clothes up, and they kept the windows open so the air would get in. Used to get some awful smells. That was in Byproducts, where they made coke for the Blast Furnace.

I still remember my last day in the mill, it was July 27, 1983. I was fifty-six years old, I took early retirement. I could've stayed but 'cause I retired, a younger guy was able to stay. I met the requirements — my age, enough time. If you were a little younger and if you worked for so many years, they offered you early retirement. If you had the age and the number of years. They paid you according to how many years you had. Which is lousy.

I'll tell you, like I said, I was in two wars and this was worse, in the mill, then the war. Every night I went to that steel mill and it was like going into another battle zone. Noise, I always wore ear plugs but a lot of the boilermakers didn't wear earplugs. Now they have hearing aids today. When I drove rivets, I always put ear plugs in. That's one thing the union did, they made them provide us with gloves and hearing aids — not hearing aids but ear plugs.

I used to have some photographs, you know I . . . you know what I did I gave them to — he use to be our president — Mike Young. They have them down community college, no Geneva. I gave him four pictures. I don't know where I got them, some kind of union thing. They had one picture of the Tin Mill and the other picture of . . . four beautiful pictures, they were in a folder too, they were great. I gave them to Mike Young, he was our president. He died but they're going to have them in that museum down Geneva College. When you go up Ambridge Boulevard. You look at that building still standing down there. I drove rivets on that building. One day we had to quit cause our hoses were freezing, it was so cold. Couldn't get air, you know, we heated the rivets in a force. We worked high up in the air a lot. Dangerous job.

Pipeshop Champion superteam

Boiler House

Blast Furnace gas line repair

Rear boiler house

Aliquippa Works
Steam Boilers

North Mill Boiler Houses
(Also Servicing Blast Furnaces)

No. & Make	Capacity	Pressure	Fuel
11 Erie City	397 BHP Ea.	150 PSI	Bl. Fce. Gas
10 Erie City	397 BHP Ea.	150 PSI	Bl. Fce. Gas
3 Babcock & Wilcox	1,475 BHP Ea.	170 PSI	Bl. Fce. Gas
3 Babcock & Wilcox	2,494 BHP (2) Ea.	435 PSI	Bl. Fce. Gas & Pulv. Coal
	2,912 BHP (1)	435 PSI	Bl. Fce. Gas & Pulv. Coal
27	20,662 BHP		

June 1960 produced 18,400-1000 BHP Hrs.

South Mill Boiler Houses

No. & Make	Capacity	Pressure	Fuel
12 Erie City	397 BHP Ea.	150 PSI	Oil
2 Babcock & Wilcox	1,530 BHP Ea.	150 PSI	Oil
5 Babcock & Wilcox	1,969 BHP Ea.	225 PSI	Pulv. Coal & B.P. Gas
19 17,669	BHP		

June 1960 produced 10,300-1000 BHP Hrs.

Reference 2

Aliquippa Works
<u>Water Pumping</u>

<u>No.</u>	<u>Type</u>	<u>GPM</u>	Type of Drive
Service Water			

<u>North Mills Pump House</u>

No.	Type	GPM	Type of Drive
1	Centrifugal	42,000	Steam Turbine
1	Centrifugal	21,000	Steam Turbine
1	Centrifugal	27,800	Electric Motor
June 1960	2,772 MM Gals.		

<u>Central Pump House</u>

No.	Type	GPM	Type of Drive
5	Centrifugal	10,400 Ea.	Electric Motor
2	Centrifugal	27,800 Ea.	Electric Motor
June 1960	2,957 MM Gals.		

<u>South Mills Pump House</u>

No.	Type	GPM	Type of Drive
1	Centrifugal	6,000	Steam Turbine
2	Centrifugal	27,800 Ea.	Electric Motor
June 1960	1,527 MM Gals.		

Total Rated Plant Service Water Capacity	225,200 GPM
Total Consumption June 1960	7,256 MM Gals.

Condenser Circulating Water

<u>South Mills Pump House</u>

No.	Type	GPM	Type of Drive
3	Centrifugal	28,000 Ea.	Electric Motor

Chapter 14

၅၁၄

Brick Layers

H e was very Latin in his mannerism, inflection of speech and dialect, and was most proud of the benefits obtained for others with silicosis. As with most bricklayers and masons the silica of sand used in mortar went on the scar and restrict the lungs making them short of breath much before their time.

My name is Albert Awad and I'm from Aliquippa. I'm sixty-seven years old and was born in Cuba. I was hired in 1951 and worked thirty-four years. Well, at that time I was a helper when I got hired. I think making about ten or 12,000 dollars a year. When I retired in 1985 — I never told this before because when I use to go to the state they always wanted to know how much you were making and I never told them the truth. I think I was making around, with overtime, around 40,000 dollars. I worked as part of General Maintenance as a bricklayer. That's correct, I was a grievance committeeman for twenty-seven years.

Well the Brick Department, what we do is tour the mill and we repair the furnaces, the Blast Furnace. We would rebuild it from the compressor up, and in Byproducts did batteries, Open Hearth, Round Mill, the ladles, just name it. So we rebricked areas and the brick was there for insulation from the hot metal. Right, and I don't know if you know the work of the Blast Furnace and all that, all that work was involved.

My first job was bricklayer helper. Yes, I was in an apprenticeship program. You worked with a more senior person, the apprenticeship lasted four years. That was not only for bricklayers, that was for carpenters, boilermakers, so on and so on. Four years, we use to go to school half a day and half a day on the job. So in 1951, I entered right in as a bricklayer's helper. I was bricklayer helper for five or six years before I entered the program. I had a little difficulty entering the program. Because at that time there were rules that you had to have your high school education, you had to be a GI, and you had to be an American citizen. I had difficulty with that, but I accomplished that . . . I was twenty-one years old when my family moved from Cuba. I married over here.

Well when I started working a rough day, when I got that job, we were rebuilding the Blast Furnace. I used to wheel brick from the boxcar to the job. Well, to be honest with you, when I was in Cuba I never worked in my life, and when I started working in the mill it was heavy. It was probably about 300 or 400 pounds and I cried like hell because the first couple of days my hands were all blisters, you know, from wheeling brick from the boxcar to the job. I'd say maybe about a block from the boxcar to the job site. All day long you take that wheelbarrow back. You had a boss watching you load and unload the wheelbarrow.

Intricate, yeah, because those brick were different shapes and molds, just like a puzzle. Lot of times those big . . . most of the times when you keyed it up, you come around like that in the Blast Furnace. You had to sketch it down, you had to mark it, and then scratch it with a hammer, or you had to send it out to somebody to cut it with the saw. So we were not laying a flat wall, we were laying inside of a round furnace. Right, and you put insulation in the back and all that stuff before you lay your brick, right. It is hard work. Some of those brick weigh over a 100 pounds and some of those were block. Well, I'll tell you the truth, the best day was when we use to go paving, they called it paving. You know, we would go re-curb the floors in the Open Hearth, they were the best days. So it was open work and the easier work, not confined like the Blast Furnace. Byproducts, that was awful. Over there we had to go . . . when the batteries got hot we had to put insulation from the heat. They didn't shut those batteries down and on both sides they're running. We repaired it and worked on it. And they don't shut down, it's continuous.

Well, you see, when I became a committeeman, I just do work, used to work probably half a day only, and half a day with the union because

I use to be the chairman of the contracting-out committee. I used to represent all the crafts, carpenters, painters, laborers, truck drivers. I used to have a lot of grievance cases and I used to have to hear with the superintendent, then it had to go to arbitration. I have to be frank with you, my best day was when I leave the mill.

Well, first of all, I know that type of work eventually is going to make you sick. Consequently, I got asbestosis — not only me — I have it, that's how I retired. We used to go for a physical, the brick gang would go for a physical every year or every two years. The doctor, well, in my case, he said, you got a couple of scars, there is nothing wrong with that. You are fit to work. And my co-workers got the same thing. My fellow workers that did pass away from silicosis — they die . . . from that and I became very interested in that. I did put some resolutions to the International Convention and I used to write to the local paper concerning that. So I was aware of the condition.

I'll be honest with you, I'm shocked that my health was affected because I was only working half of the time, half of the time I was working out of the mill. Can you imagine those people that working full-time, day by day? I remember all of it. I remember all of them, I was their grievance man. We worked together, we had parties, it was like a family.

No, they didn't acknowledge any responsibility. The sad part about this situation is that I think the people became aware through my articles not the company and people tried to help in the union. You know, we have a couple of lawyers from Pittsburgh that were concerned about us. We established a hotline and a paper and we had this lawyer, who worked on commission. They wouldn't charge you nothing, and then they started bringing in trailers. The union was examining the guys and took x-rays and the ball started rolling, and then we filed a grievance on their behalf.

That was awful, the effect on my family, yeah. Tell you the truth, I don't remember raising my kids because I came over here and I didn't have no money. I started working overtime and then with my union job . . . so I was supporting my family, building a house, and the time flew.

Cuba was beautiful, but not like what the people picture. You have to work hard over here. If you are aggressive like any place else, you know you get something, you get the opportunity. But it's not like a lot of people think, and its not like a lot of people think, and this side of the ocean after you reach certain point you have to keep working just to maintain that.

Yeah, even over here the people think the same way, they see that you have something more than them and they think somebody gave it to you — you didn't work for it.

My last day in the mill, well, I got this ability to file a case against the corporation and I got my social security. I got my workman compensation and I can't complain, frankly, I can't complain.

You don't have different philosophy . . . a lot of people here, a lot of people talking in this country. My only view is different completely. I have two boys and two girls and I always taught them that nobody is going to give nothing, and you have to be aggressive. You can't be timid and never take no for an answer. I always taught them I don't want to hear who told you this and who told you that. I want you to explore the area, listen, don't argue, but don't trust what they tell you. That's my philosophy.

Unfortunately, I am one of those guys, when I was in the union, I used go to negotiation with the company and used to go here and used go there. I was an inside information, told people they were going to shut it down, but people didn't believe me. I mean, people, they don't know how to draw the line even with the union. They always want more. So they priced themselves out of work.

Furnace lining

Brick and scaffold

Oh yeah, I do remember very well when they started shutting down the mill little by little, you know. Senator Heinz came down and I was part of the tour — the union, and the company, and Senator Heinz, and everybody starting questioning. I just listened to what they had to ask and then when the tour was over we went to the union hall. I say, senator, I would like to ask you a question. He said, yes sir. I say, my company tell me that we can't survive, they are losing money, my union tell me the same thing.

Let me tell you the truth, he said, you know, if you guys, all of you donate your wages for a year this place would still go under.

Yeah, it didn't matter. He said we can't compete with the Japanese, we compete with this or that. I said, well, if it's true that if they buy the steel cheaper, how come they don't pass the fruit to the workers, why they keep all the profit? They are making triple the amount of money. He said, Albert, that is business.

Base of Blast Furnace lining

Chapter 15

ℰℛ

Carpenter Shop

H e was balding and a roundish man, in a nylon baseball jacket, sleeves pulled up like a scholastic baseball coach.

My name is Carl Ross and I'm sixty-five years old. I was born in Aliquippa, 1827 Davison Street, upstairs.

I was in from 1948 until 1985 for thirty-eight years and eight months. Call it retire, kind of forced out in 1985. Well, I was on salary and I guess it was like 2,500 dollars a month. When I started, I was making about a dollar forty cents an hour. The final years I was making over 30,000 dollars. I worked in the Carpenter Shop, it was called the Service Department.

The Service Department took in brick layers, the garage, the mechanics, the laborers, the Tin Shop, the Carpenter Shop. That was all Service Department and they serviced the other crafts. They serviced brick — well, they did work for the brick layers — but they serviced the boilmakers, the riggers, the pipefitters, and all the other craftsmen.

When I first started I was a saw grinder. I was in the Welded Tube Department and I worked from 1948 to 1951. Wait a minute, let's see, I was in the Army when I was in the Carpenter Shop, and that was in 1952. Yeah, so that would be 1948, '49, '50. I was a saw grinder and

Carl Ross

I was in the Machine Shop down the Welded Tube. The saws they used to cut the hot pipe. That was what I used to sharpen and that was the first job.

They had a machine, an automatic machine, it pushed the tooth underneath the wheel that went up and down. The emery wheel that went up and down, and it just pushed it through and the wheel come down like that. They had three machines, so you run those three machines and sharpened the saws. That lasted until 1950, and then in '51 I became an apprentice carpenter.

Well, I applied for the apprenticeship system. They gave you choices and my first choice was electrician, and then carpenter, then bricklaying. So when it came through they gave me the carpentry. So anyway, you go to school, you were supposed to go a percentage of your working hours, you were supposed to go to school like four hours for a forty hour week. You had to go to school, and at the end of school there's an ICS program. The International Correspondence School, and then you go and you'd have your books. You were given assigned books and at the end of the book would be a test. Then you would finish your test up and the guy that was in charge would send those into the ICS headquarters. Then they would grade them and then send them back. So this is how we went to school.

We had other forms that we had to fill out that said that you performed so many hours scaffold work, so many hours form work, so many hours different work. They would have these cards so often and you would have to sign them. So sometimes there were a little bit of discrepancies on the cards. They just filled them in, but you could've worked for scaffolds like, all week, and they would put so many hours scaffold and so many of this.

A couple of us got to wondering, we wanted some shop work. We wanted to learn carpentry and we wanted to learn how to make cabinets and different things like that. They had just a special crew that did that in the shop, so we wanted some of that time, too. So we refused to sign some of these cards. So then they called us in one by one and more or less told us either sign the cards, or . . . so anyway.

Scaffolding was the most prevalent operation. We would hang scaffolds. The Carpenter Shop did as much if not more climbing than the so called riggers. We would go first and we would hang a scaffold, and then the riggers would go up on our scaffold and work. Yeah, we would put the scaffolds up first, and not too many people know this about carpenters down the mill, that they climbed just like the riggers. But if there was a boilermaker and if he was working on something and he needed a scaffold, we would put it up. If it was what we called the Bull Gang — the Bull Gang worked on cranes and underneath the hoist there was a drum and the cables had to have new shims or small supports used to level large pieces of machinery put on. Well, we would hang a scaffold underneath the crane up in the air. They would go down and they would work on their drum, put new cables on or whatever they had to do, and we put those up.

In Byproducts, we put scaffolds out on the riverbank where they would unload the coal barges. Those shovels would continually go around, so we would have to hang scaffolds out there so that the boilermakers could fix or repair the buckets, and the riggers would work out there, too, and then we would hang scaffolds for them. So the majority of it was scaffolds, and we had a lot of form work — foundations for crane runways, retainer walls in the Byproducts that we put up, and any new building that went up any new installation of new equipment. We had to build a forms and they poured the concrete. The laborers would be there to pour the concrete and then we would stay and help them in case anything broke loose or something like that. We did our own what we called rebar work. All the reinforcing rods, well, we put the rebars in. We did all that and then we put the forms around the rebars.

And then furnaces, Open-Hearth furnaces, Soaking Pits, the Seamless Tube, Welded Tube furnaces — all the furnaces when they were relined or rebuilt. They would bust the brick down and the laborers would go in and clean out the brick and, well, the base was still hot. We used to make wooden shoes to put on their feet. So they wouldn't burn up and then we would put our shoes on so we wouldn't burn up, and go in and start putting in.

If it was an arch roof, we would have to put in an arch roof, and the biggest roofs that we put in were in the Open Hearth furnaces. The open-furnaces stretch, oh, maybe about between twenty-five to thirty feet across, and had a rise on the roof about maybe twenty-four to thirty inches high. That was the arch, so we would go in while it was still hot, they would knock the brick down. But the hearth was still. They didn't take nothing out of the hearth because it was molded metal.

It was still hot and you would sweat your rear-end off, but anyway, we would go and then we would put asbestos down and put timber on top of this asbestos so it wouldn't burn. Then we would put posts underneath these timbers once the big timbers were in, which were eight-by-ten's. We put post underneath and then underneath those post and then under those posts we had to put asbestos on the bottom. Then, once that was down, then we laid four-by-six's on top of those timbers and, then, once those were on, then we put the centers, what we called centers, on top of that. Then the lagging or two-by-four's were nailed to those centers. It made a complete arch so the bricklayers just laid their brick right on our arch, and we had wedges underneath our centers that raised them up. So once the roof was arched and we knocked our wedges out, this would drop down. We would have to clean that out and it got a little warm in there. It was a sweaty, sweaty job, a lot of the furnace jobs were sweaty.

We had shop guys for fine work such as cabinetry, but they were in the shop constantly. Then we had a pattern maker who made patterns. They would take patterns up to the foundry in the Blast Furnace and they would lay these patterns in a loom and pack it and surround it with sand. Then they would take that out and then pour the molten iron in. This was cast iron, it wasn't cast steel, it was cast iron. Cause the only thing that came out of the Blast Furnace was iron and anything that came out of the Open Hearth was steel, and anything that came out of the BOF Furnaces was steel.

BOF Furnaces were another thing. Blast Furnaces were a big thing, it took us months to rebuild the Blast Furnace. That was a tremendous job. When they relined the Blast Furnace they began at the opening at the top. The riggers would take out all the brick at top and there was a big hole, a big circle, maybe about thirty feet this diameter here, right inside here. We used to make a scaffold and it hung on four corners and they had hoists, air hoists to support the workers. These air hoists were hooked up to this scaffold, well underneath these scaffold we made a solid, flat point where they joined. But we had big U bolts that went underneath, to hold everything together and we had timbers eight-by-eight's or eight-by-ten's — that we used to add for strength since the furnace was built on a taper. So as they came down the laborers would bust all the old brick and let fall to the bottom, and the furnace got wider as they went.

We would go down and we pulled those timbers out and they called them outriggers, and then we put plank on them. Then they got closer to the wall and they were able to bust those bricks and let them break it down. So we go down, and then to take the scaffold out we would bring those in and then they would take the scaffold out. This was where we lost one foreman, down that furnace, when they picked it up. When it got up to the top there are floating timbers underneath and we used to wire them up, so when they bring them over to the edge they wouldn't slide out. Well, what happened was they had it up and this foreman was there and he was going to help out. He got underneath there and just at that time the hook tilted and tilted the platform. The timber slid out and hit him and drove right down to about 100 feet to the bottom of the furnace. He got killed, and then it pinned another one up against the edge, it pinned his leg. So we had one guy hurt and the other one got killed, but there are several stories, and then there are other stories.

We had guys fall into a big . . . we called it the mother liquor tanks. Did you ever see them big oil tanks? Well, these were the kind they had over in the Byproducts. Up on top there was hole in the roof of one of these and somebody had thrown some asbestos cloth over top of the hole. Well when we were up there and we were going to put scaffolds and plank to walk on, this carpenter helper happened to walk over and he hit that asbestos cloth and he went down. That was a horrible death. Jughead Walker was his name, that was a terrible thing. Anyway, that was some of the jobs that we did.

But the Blast Furnace, once we got that hanging scaffold out. Then we would start on the bottom and we started out with twelve-by-twelve timbers and put them up and put a base platform in what we called the casthouse floor. In the casthouse they would feed all their brick into this platform, and as we went up with our scaffold every five feet, we would go up five feet and put a deck. Then when we would go up, we would take that deck off and put posts on and then move that deck up. The next time we would leave that deck and start a new deck. Every other deck was new.

It was being with the people, being with the guys. That was . . . General Maintenance was like a family. In this family, say you're a rigger and you needed something. The carpenters could fix you up. You would go over to the Carpenter Department and you would say, I need this, can you help me out, and they would do it. The riggers would do the same thing, the boilermakers did the same thing, the pipefitters. If they can help you out, they would, and the whole General Maintenance Department was like that. Just one big family and they got along very well, and if you would ask all the guys, do you miss the work? No, I miss the guys. This is about what you would get.

Now in the Carpenter Shop we still have our annual picnic, and I think the riggers still have theirs, and I think the bricklayers and the pipefitters, most of the departments still meet yearly. The thing that I would like to forget is the deaths, things like that, those were the bad things. We had a lot deaths down there. It was due to the jobs.

They had a crane and they were putting on a walkway. They were rebuilding this walkway on the crane and it was dark. We were on four to midnight and it was a rainy day. Maybe it was a drizzly day, you could see it was night time, but it was like five o'clock. You could still see, but up in the air when you got up there you couldn't see underneath. When he was cutting off this walkway, he cut this one walkway off, and underneath, the bolts weren't in where he was. When he cut this one section, it was holding up the section that he was on and it came down, and he came down.

I can still picture it today. I don't know how you feel about me telling you about it. Anyway there were railroad tracks and he came down on the one railroad track on his head, and his head, you can see it was open and blood was coming right out. Mitchie's holding him and holding him. So the day was bad, it was drizzling rain, it just put a damper on the job. But when that happened everybody . . . we shut it down and everybody waited for the ambulance and when it came. And then everybody went home.

Well, my first born, we would work twelve hours on and twelve off so when I came home she was asleep, when I left she was asleep. So she didn't see me for maybe a couple of weeks, and then when I went to pick her up she cried because she didn't know me. I really felt bad, I said, geez my own kid, and she treats me like a stranger. But that was the way it was. But the thing about the mill was that you spend more time with the guys in the mill than you did with your own family, 'cause you would be there.

Say you were working twelve on and twelve off, well, you're with your guys half hour before the shift, half hour maybe after. You go home, you're only home for a few hours, and then you go to bed. So you were up more with your friends or the guys you work with, then you were with your family. Even if you worked eight hours in the mill you were with your guys, if you came in early, half hour early, so you had eight and a half hours with them. You go home. How long do you stay up, and then you go to bed. So actually you spend more time in the mill with the guys you work with than you do with your family.

We use to have different guys, like the riggers, Red Patterson would be a name to throw in, and this Francis Zitzman was General Maintenance boss for us. They use to come around when we were rebuilding the furnaces, so they used to come around on the jobs and I would see Red. Red would go sometimes, get the hell out of here, we know what we're doing. And he would tell him.

They understood him and they knew he knew his job, and they didn't give him no back talk. We had what we called high flyers, good climbers. We had guys by the name of George Blinky and Joe Letteri, Mario Letteri, Jack Cable, the one that just died. These guys were high flyers. They were good in the air, they were good climbers, good scaffold builders in the air.

We'd be anywhere from . . . you start from the ground level and the highest was the Blast Furnace preheat. The preheat was over 100 feet high, and then we would have to hang scaffolds out over the preheat, and then even on the blast furnace. My fingerprints are still in that piece of metal up there. I think they are still in that piece of metal up there at the top of the Blast Furnace.

The riggers use to have a Shiv wheel and a Shiv wheel, would carry the cable, and it brought everything up to the top of the furnace. Well, out there they had two beams going up and there was like two little ladders coming down, one on each side of each beam. Well, we had to go out and put plank between those ladders and we had to wire that down. Then the rigger could get out there and sit on that plank and then feed the cable up to that shivel. So a friend of mine, George Skorich, we went up one time and we had to put that plank up. When you looked, and you looked straight down, that's about 100 feet, and all you see is space. That never bothered me, because I always said if something happens I can grab this, I can go there. But there, there was nothing.

These were times when you didn't wear your safety belts. Safety belts were cumbersome actually, sometimes they hindered you more than they helped you. Still, we didn't wear them, maybe we should've more, but that's the way we got started. You didn't think of that, anyway, this is some of the things that we did.

Yes, a lot of things could've been changed, power equipment. See, when you're a carpenter you had to pick up a sixteen foot plank and pull it with a rope. Sometimes you pulled that thing twenty, thirty feet high, and you're doing this manually. Now one sixteen foot plank weighs about eighty pounds, provided it's not wet. If they're soaked they weigh like a ton, but anyway, I would see a big crane and they would pick up a little angle iron that you could pick up with one hand and carry it up. But this is what they did, they used the crane to pick up metal. When it comes to wood, we were secondary. You're a carpenter, this is metal stuff, we had to take a backseat. So that would be one of the things that I would change, would be to get some equipment to help pick it up.

So finally, in the later days, we even had one big helicopter come. We made a landing platform for him for this helicopter to drop the lumber because we had so much work to do on the roofs. They were working on power lines, electricians were working on these tires to prevent electrocution. We had to build scaffolds on all these tires that went along the mill, so to get the lumber up we used to punch a hole in the roof. The truck would come down below and we would have to pull all this lumber up by hand. Then when you put it all up by hand and you carry it along the roof tops, you build your scaffolds.

So that might be one thing, to get power to help the guys get the material to the job rather than them doing back-breaking work. That would be the biggest thing, and some of the other things would be some of the equipment, the power tools, nailers, and chain saws, and things like that weren't given to us until in the later years. They were out, but

our bosses never bought them, so we still had to cut lumber and everything by hand. When we go in the Blast Furnace, then they would give us the portable saws, but on regular scaffolds they won't do it, so, anyway . . . Eventually it got around and we started to get more power tools, but it was too late then, this place started to fold up.

Erecting work scaffold

I really don't know, why is Cleveland working and we're not, why is their mill still working? A lot of people tell their kids, oh, you don't want to work in the mill, but that was because the mill had some hot, filthy jobs. But if you would look at it pragmatically, here is a job, here's the way to build your house, here is a way to send your kids to school.

I got my house paid for and I sent my kids to college, and I did that through the mill. A lot guys outside the mill that went into business on their own, they always complain and say, oh, you got a pension coming. I said that was your decision, you wanted to go into business, you didn't want to have a boss, you wanted to be your own boss, now you took that road. I took this road, now you're begrudging me because I took that road. So, one way or the other . . .

Scaffold for Blast Furnace lining

The mill, it helped a lot of people in Aliquippa, and even though they say I don't want to get into that area, they made the same money as we did. Everybody, and without it you can see what's happening — the town is just falling apart. It was a booming town at one time, it's still a nice town. The people in it haven't changed, some of the kids have changed, but mostly the old timers are there. They still . . . guys still come over the house if they need something. Can I help you out, that part is still there. The friendship is still there, but it was a good job. Hard work, but it was a good living.

Chapter 16

ℰᴕᏩᎡ

Electricians

H e was quite distinguished, a beaming smile, the man who preserved a wealth of pictorial and factual information concerning J&L. As the dismantlers tore the mill down to be sold for scrap, he literally dragged cabinets of photos from the mud, hauling them out on his own, while others watched.

My name is Don Inman and I'm from Beaver Falls. I am fifty-five years old and was born in Patterson Township, which is Beaver Falls.

I have a little over thirty-six, of service. I was hired on January 18, 1960. In 1960, I made a little less than 4,000 dollars. Now I'm making 50,000 dollars, but it's because of the position that I have, too. I started out in the Strip Mill. I started out, of course, worked as a laborer as everyone had to. Went into the Strip Mill Electrical reasonably quick, did my training in the Strip Mill Electrical. They, of course, had, or were trying to establish, the apprenticeship program. I did studying on my own and tested out of the total apprenticeship program, and was able to get A-rate. There were some discrepancies during that period of time. They tried to regress some of us that had done that. Very few of us had, but they tried to regress some of us that had done that. We were able to file a grievance, that's the only time I filed one. Go to the company and

they said, well, you'll have to prove that you had the time. I said, well I do, and they said, how many hours do you have? I said, about 5,050, is that enough working at A-rate. So they checked and that was very close to what I had. So then I was given the A-rate, when I worked it, even though I worked, as they classified, as a helper as well.

Donald Inman

I was in the Strip Mill for about fourteen years and then, when they put the bids up for the General Maintenance Electronics . . . They put it up and I was fifty-fifth on the list, and they ended up taking five and I was number five. Four people passed the exams before I did. I was very fortunate, there were nine elements of the exam, I passed eight of the nine and therefore I was able to get what they called B-rate. So I worked at that rate for six months after the exam. I took the exam again and then went on to A-rate.

And then, at that point in time, they came along again and decided to change the job description. They changed it again and decided to give us the job description of Master Electrician. Well, because I had passed the different tests for motor inspector and for the different things, all I had to do to get the title Master Electrician was go to welding school, because I had the electronics apprenticeship. I had all the other ones, so I went to welding school. Then they decided that I had to retake high voltage school, which I did, and then was given the job description of Master Electrician. So that's actually what my title is, originally it was the Electrical Department within the 44 Inch Hot Strip Mill.

The Electrical Department was within General Maintenance. Well, the section I was in was electronics, so we were attached to the line and wire gang. So we went out in the mill and we took care of the radio controlled locomotives, the Blast Furnace, and the BOF. We took care of the communications of the BOF and the Strandcast and the Blooming Mill and the lower level, the 14 Inch Mill. We took care of the electronic scales that the Blast Furnace and the BOF used. We took care of the actuary gauge in the Strip Mill. I left the Strip Mill.

They didn't have anyone to take care of it, so then I went back to the Strip Mill and took care of the actuary gauge for them as well. Then, when I worked in that particular group, also took care of all the two-way radios in the mill. We had 365 two-way radios that I took care of by myself.

I don't know, there was about six or seven of us in electronics. I'm not sure how many there would of been in line and wire total. Well, in the heyday there was a lot, there was a couple of hundred in the heyday. But I mean, when I was there, I suppose there were maybe eighty or ninety at that time.

Well, my day was a little bit different than most peoples' because I went in and we had a wash room right next to the electronics shop. Therefore, I would go in and start on my two-way radios. I would work virtually all day on those radios because I was responsible for just those where everyone else was responsible for everything else. However, we all took our turns working four to midnight and when I had to take my turn working four to midnight, I had to work by myself. So I had to be familiar with the electronic scales and all those things I talked about before, because there was no foreman. You were there on your own. Therefore, if I could get caught up on my work, what I would do then is I would try and go out with the fellows and go maybe to the Byproducts and run some of the lines with them. Find out where some of that stuff was because, as I said, when you are on four to midnight, you're there by yourself. You really had to know where that stuff was.

I had a lot of good days, I really did. I suppose maybe not the best day, but one of the most interesting days was the first week that I went to the line and wire gang. I went as one of the five and all of us went from the Strip Mill. I was asked to go work on the two-way radios. The reason I was asked, I was the only one who had their Federal license. When I had gone to electronic school I went ahead — it was in Washington, D.C. — and I had gone over the FCC and then taken all the licensing that they offered, so I had the Federal license. So when Joe Burns, who was in charge of electronics, asked me to go work on this equipment, I went and worked on it. They were just everywhere.

I couldn't get over how much equipment was around that needed repairs. So I started working on it, I worked like crazy on this stuff. I felt bad because I wasn't getting what I felt was enough accomplished, and so at the end of the week I think I might've fixed fifteen or sixteen radios. I felt bad because I wasn't familiar with the prints. I wasn't familiar with the radios, and I had to go through all that and learn this. Well, anyway, so the week went by, and Monday morning I came out and the fellow that had been on vacation that I had replaced for the week was sitting where I had been sitting. He had his head propped up on the chair.

I went over and — I'm working on a trolley phone with a friend of mine, no use of getting into this. Joe Burns come back and he looked at me and he said, what are you doing down here. I thought, oh my, I've only been here a week and I'm in trouble. I said, well, I'm trying to repair this trolley phone with Pete. He said, why aren't you up there fixing radios?. I said, because it's not my job, it's here, and this gentleman was over in the corner and was asleep. It was about seven-thirty a.m. and he's sleeping already.

He woke him up and told him to hit the road, that I was doing the radios. Well, of course, you can imagine the shock. He was just perfectly still and I said, well, what choice do I have, so I started to work on the radios. Well, what was interesting about it is I had fixed fifteen or sixteen radios. I didn't realize that this fellow was getting one every other day. So it kind of made my day in a way, cause I didn't feel quite so bad, of course. Then when the mill worked, it really blossomed. At that time he had maybe forty or fifty radios in the mill and, of course, it really blossomed, and like I said, we had between 300 and 400. It was a real challenge to keep them going, plus the repeaters and the base stations and things like that had to be done.

My worst day, which was not a bad day, my worst day was a twelve to eight. I was still back in the Strip Mill and when I started working in the mill I was 117 pounds and was five foot eleven inches. So obviously I was very thin. They had a set of edger motors up on top of the Reversing Mill, and they were very close to each other — probably maybe a foot apart and no one could check the brushes. Well, I had no problem, I would just slide down between the two motors and I would take the covers off before I did. I would stick my hands in the motors, I would change the brushes, and had no trouble whatsoever. Anyway I must've been there eight or ten years and this midnight someone had to change the brushes. No problem, right.

Only, I probably weighed more, probably fifty pounds more than I did it the first time, and got down in there. I was changing the brushes and realized that I couldn't get out, and that I'm stuck down in between the motors. It's night turn, midnight, and I'm not totally alone, there are two or three other fellows out there somewhere, but who knows where. Never had claustrophobia until then, and I realized I was truly in trouble. I was probably down in there for about an hour, and fortunately a friend of mine was expecting me to come back for lunch. I didn't and he come up and helped pull. I was in there in such a way that was no way I could get my shoulders out of the motor in order to be able to push myself up out. When I was younger and thinner that wasn't a problem. So that was probably one of my worst days.

Well, it was a good place to work. It was a good living. I enjoyed my days in the mill, I can't say that I'm really one to regret it, going to work. I always thought it was a nice, honest days work. I loved the people, loved the mix of people, because everyone was different, everyone had a story. Everyone had a different perspective and one of the reasons that, I think . . . Well, one of the reasons I know that Aliquippa survived so well was because of the mix of nationalities and people. Because when you would work on something you'd have an idea of a way to do it. There would be somebody standing next to you who had a whole different perspective and he would suggest something else, and you would have four or five people, and before you knew, you were doing altogether different than any one of the five would have done, but it worked. Good experience, it really was.

Well, not really. I would say some of the things that affected me though were some of the deaths in the mill and what not. I remember being on the communications detail, when out at the BOF they had a fellow slip on one of the platforms and went into the ladle. One of the things I had to go out immediately, they had a communications system from the crane down to the platform where they talked to the craneman, and we had to go out right away and check that system, and they hadn't taken the ladle out yet. That was a little unnerving, knowing that had happened and you're standing on the expect spot that he was, and that kind of thing. And then they just took and poured a little bit of the ladle out on the ground, and that was the body, and the rest they just went and used.

Well, oh yes, there is a lot of people that come to mind. One of the fellows that always kind of amazed me and impressed me was a gentleman by the name of Joe Giammaria. He was in charge of the electrical in the 44 Inch Hot Strip Mill. After I was hired in the Strip Mill, of course, but I . . . everyone is hired in as a laborer. I went to him and I told him my qualifications, what I wanted to do, and he said well, if I get any openings I'll let you know. Well, that's what they all say, right. But I didn't bother him at all. I worked out of the labor gang, we all had different jobs. I did a lot of different things around the mill and what not.

But anyway, about a year and a half past and one day he sent down for me. I came up and he said about a year and a half ago you talked to me and wanted to know if you could have a job, and he said, are you still interested. I said, yes, and he said, well, how about coming up here on Monday and starting. I was very impressed because he seemed to be a very fair person, a very knowledgeable person, and he remembered all that. Those things have a tendency to stay with you, and you're impressed with people like that.

There are other people. We had a fellow who is no longer with LTV, he was in the Wire Gang, too. We called him Rabbi, of course, everybody had nicknames. Rabbi came up from the Wire Gang. They would bring people up from the Wire Gang to work with us on temporary assignment and what not. I was always impressed with how intelligent he was. I could tell him how I would fix a radio and he had no background, but yet he would just grab it right away and go fix things that I had worked on. We would go out in the mill and show him things. It was unbelievable how quickly he would pick it up. I guess it depends on what point in time.

I guess in the very beginning, well, my dad started in 1917, he worked in the mill. Between him and my brothers and uncles and what not, there has been Inmans in the mill since 1917. I'm the last one, I guess, a dying breed. Anyway, I remember as a child my dad was working in the pay office. I remember him coming home and telling different stories, an example, Brinks went on strike. I remember him going to Pittsburgh and counting out. At the time he had to count out three million dollars and to me, as a child, that just seemed like so much money to entrust it to one person. He told me about them counting it. How they had guards around him while they were doing all this because, of course, they paid by cash in the early days.

I remember this story that he told me was interesting. One of the first jobs that he had was working the pay wagon. I don't know if you've seen any of the pictures of the early pay wagon, but it was a wagon pulled by horses and it had round windows on the side. They would take the wagon out into the mill and then the people would go there and get their pay and, course, they paid by cash. Probably, he started in 1917, probably 1919, or somewhere in that period of time. He was on the pay wagon, he and his boss. They had taken the wagon down to the Tin Mill and they were paying the Tin Mill, and anyway, they were finished paying. They came down with the horses, hooked the horses up, and were bringing the wagon around. The wagon turned and one of the wheels slipped down into the track. So his boss said, well, I'll go out and tell them our problem here and I'll be right back.

So he came back and he's sitting in the wagon, and my dad is sitting in the wagon and off in the distance he hears this siren. It's one of these wind up sirens. So he said to his boss, I wonder if there's something going on in the mill, and he said, well, I don't know. When I called up plant protection, he said, I told them we were held up down here, and it would be a little while before we get up there, he said, I don't know what. It couldn't be us, he said, as soon as he heard what he had said he realized that they were in trouble, and so they sat in the wagon and here comes the Aliquippa Police with the J&L guards, and one of them had a tommygun and the whole nine yards. They came down, surrounded the wagon and everything, and they had to talk their way out of what had happened. So that was kind of interesting.

I remember as a child, too, when my dad would work over when they had to count money and what not. They paid salary with a two dollar bill. Of course, I was younger, I was the youngest one in my family. My dad was forty years old when I was born, so he was probably forty-eight or fifty when he started doing that. But anyway, he always give me the two dollar bill. He would always go out to lunch and spend his own money and saw that I got the two dollar bill. It was a good life, the mill was good to my family, put my brother through Carnegie Mellon, it certainly raised us. Here I am with a home that is certainly adequate and what not for my needs, so it was a good experience.

Well, that's a good question. I have often thought that I made a mistake by not getting a better education, taking advantage of the opportunities that were available to me, but I don't know. In retrospect, I'm not so sure that I would change much. I was offered salary three times after I left the Strip Mill. They wanted me to go back to the Strip Mill as an electrical turn foreman. While I was in General Maintenance they wanted me to go to the Blast Furnace as an electrical turn foreman. They asked me to go to the Wire Mill as an electrical turn foreman, and I went on the interviews and listened to what they had to say and turned each one down. The Wire Mill is gone, the Blast Furnace is gone, the Strip Mill is gone, and I was in a roundabout way offered a salary position in the Tin Mill. I told the gentleman who offered me that, well, really the job that I have is probably as good as it gets.

Well, I'm sure for them it was similar to mine. In 1982, I was laid off and I was laid off for nine months, and it never occurred to me ever that I wouldn't go back to work. I think that was part of the problem for many of the people that worked in the mill, it had always been there and it always would be, and I don't think anyone really thought that they would ever tear it down. Now, later on I think they saw the handwriting on the wall. But until the Blast Furnace had really starting to come down I don't think people had realized that that was what was going to happen.

I realized, probably earlier than most, because that's why I started going around collecting things and trying to get permission to collect things. I started taking photographs and asking people for photographs and trying to collect memorabilia and things like that.

Well, to try and preserve what I could. It seemed like no one else was and at the time I was really too old for trying to collect a lot of the stuff. In fact, some of the patterns that we have today at Geneva, and some of the material in the mill, but I went up and was given permission to take those out of the pattern storage building I couldn't even get the people that were loading them on the trucks to help me. I went up and loaded a large dump truck totally by myself, while I watched seven or eight fellows that were paid by LTV just to watch me load it. So, there is a lot of resentment, a lot of resentment from those people because they were still behind tearing the mill down, and even some of them were my friends and I think there was a lot of resentment.

Nobody ever thought it would happen. When you take like my family and other families, some of their parents and relatives worked through The Depression. One of the things that killed the mill was the fact — or at least at Aliquippa — was the fact that they put all their money into the Seamless. If the people would've been aware of what Seamless was over the years and how it was a very cyclical business, they would've realized that was not the thing to do. Because during The Depression the Rod and Wire Mill is what kept Aliquippa going. I don't know if you're aware of this or not, but things that kept it going was coat hangers. People couldn't afford to buy a lot of things, but people seemingly bought coat hangers or at least they came on clothes or whatever. Aliquippa produced the wire for coat hangers.

Whenever they shut the Wire Mill down I thought it was such a shame, because they were going fifteen dollars a ton for wire for coat hangers and things like that, and they were getting 845 dollars a ton for seamless pipe. Well, the thing that the accountants felt, they put all their money where they could get the most returns. So they put it in the Seamless, not realizing that even the Wire Mill was still there. Today, they probably would still be getting fifteen dollars a ton, that's not a big return but it's a return. It's not a lot and J&L looked at it differently than LTV. The Wire Mill shut down sometime in the 70's, but I would have to look it up to be sure.

Well, it was a good place to work, it was like a city. At one time there were 15,000 people there, so obviously larger than the city. We had two ambulances, we had two firetrucks, you had everything that a small city would have, and it was a good place. It was just a shame that the people didn't have a better education and better training, so when it did go down they had other opportunities.

That was the biggest problem, was when I was trying to learn electrical and do some of these things, a lot of people ridiculed me because they thought that they could make more money in production. As it turned out, my job skills were useful and transferable, but if you were a coiler operator, or a Finishing Mill operator — or some of these jobs required a lot of them — but once the Strip Mill was shut down or once the Blooming Mill was shut down, who wants a roller? So it was important that they get a good education and most of the fellows that went into the mill didn't do that.

Electricians shop

He was one of the lucky ones, as the mill was dismantled around them, the Tin Mill continued to work. They were fortunate and knew it, but tentative and gun-shy as their friends and neighbors lost the jobs they still had.

My name is Lelio Iannessa and I'm from Baden. I'll be fifty-seven. Believe it or not, I was born in Italy, the Province of L'aquilli. I worked forty years and was hired in 1956, July 3rd. I was making two dollars and fifty cents an hour in 1956. I didn't retire, I'm still working and making fifteen dollars and ninety-four cents a hour.

I work at LTV now. LTV is running currently the Tin Mill. I am an electrician there. Well, basically I started out as a helper. My job description at that time was basically to help the regular electrician with whatever job they had to do — change the motor, change the cables, run crane, whatever had to be done. It has changed up until this point, where I am an A-rate motor inspector electrician. Presently I work strictly out of what they call three tin line. Three tin line recently had an upgrade where ninety-nine percent of it is run by computer. That's my job, to make sure that everything . . . that the computer is out on the job.

Back, way back then they didn't have apprenticeship programs. Basically, you moved up through ranks. As you become older, when it was your turn, they ask if you wanted to go to the electrical department. You had to put in X amount of hours. You started, out as I said, you started out as just a helper. Then you went from a helper after so many hours to a C rate, to a B, to a A, but in the meantime as you progressed. In order to advance you had to take X amount of tests to make sure you knew all the qualifications and you were able to do the job. Believe it or not I was able to go from a helper to an A rate with five and a half months.

A typical day for an electrician is you basically, when you come in, your foreman lines you up or whatever has to be done. First thing, you go out on the job that you are working and you make routines, make sure everything is working right. Anything that breaks down is your responsibility. You have to find the fault, you have to find the problem with it. If need be, most of the time you do it yourself, but if need be you call for some kind of help to help you. Say if you got to change your motor or something big happens where you do need help. But most of the time it's jobs that you can handle yourself, and it's a one man operation. You just go from there. The idea is to try to keep a minimum amount running so that the line does not stop.

That's hard to say. Probably my best day was when I actually got to see the line running at full speed. When I am saying to you full speed, usually we run about 1850 feet in seconds. My best day probably is when that line is running for 1850 feet nonstop for a full eight hours.

Once again, you know, there is so much that happens here that to cut it into individual days is pretty hard. Probably the worst day that you can have is one of your major motors blows up, grounds, shorts, whatever, puts it out of operation. This just happened, as a matter of fact, last week. Where our main generator went to ground and the line stopped automatically. So now you got to go down and try to find the problem, and once you find what the problem is, try to correct it. As I said before, to change the main generator is not a one hour or two hour job, you are looking at probably ten to twelve hours. That means you got to have that line for that amount of time, and to me you're not making any production. When you're not making any production, to me it's a zero day.

The things that I remember the most is the first day that I got hired. I was hired in as a craneman, although everything was so new and so strange. I remember going up on the crane and just watching the operator that was running that crane at that time run that crane, and I thought to myself: I'm going to get paid for doing this job, this sounds pretty easy. It wasn't easy. Once you become familiar with it, you find out that you have a lot of responsibilities.

The thing that I would like to forgot is that I've seen more than one person die on the job. For whatever reason that may be — usually negligence is one of the biggest — but I've seen a few people die on the job. That is the thing that I would like to forget when I think back. I'd say a little bit of responsibility on both sides — both company and worker — but sometimes we, as human beings, take too much for granted.

I think the biggest effect that the mill had on my family was back in the early 80's or mid 80's when they started shutting departments down. Although I was one of the fortunate ones — as I said I'm still working — I also went through a layoff of eighteen months and that was quite a trauma. Didn't know if I was going to get back or what the score was.

No, I would not, and I'll tell you why. I enjoy doing my job. It's challenging, to me it's something new every day. It's not repetitious. To me, I like doing what I'm doing and no, I wouldn't change anything. But compared to when I first started, well, that's a horse of a different color. When I first started in the electrician gang, I think we had 117 back in those days cause I just looked it up at home today, we had 117 people.

That's for the whole department, one department. Today we have thirty-six people doing probably more work than what they used to do before. Times have changed because the machinery has changed, and yeah, we are down to a very minimal amount of people right now. The quality of the work is still good, believe or not, productivity is up. Once again, I relate that to the new type of machinery that they have compared to the old type. Well, worker efficiency does have a part in it, but the biggest part as I said is with the machines, the new type of machines that they're bringing in.

What I would tell a young person today is forget the mill. First of all, I don't believe that there is a future for everybody in the mill. There just aren't enough jobs. I would tell them, if at all possible, stay in school, try to get a good education, and work to stay away from there.

Electrical power supply

System Orientation

The Mill

Aliquippa Works
Electric Power Generating Facilities

North Powerhouse

| 1 | 20,000 KW | 6,600V AC 25 Cycle Turbogenerator |
| 1 | 1,500 KW | 250V DC 60 Cycle Ignitron Rectifier |

South Powerhouse

1	17,000 KW	6,600V 25 Cycle Turbogenerator
1	10,000 KW	6,600V 25 Cycle Turbogenerator
1	25,000 KW	60 Cycle to 25 Cycle Frequency Changer

14" Mill Motor Room

| 1 | 10,000 KW | 25 Cycle to 60 Cycle or |
| | | 60 Cycle to 25 Cycle Frequency Changer |

Total Aliquippa Works – June 1960 Electric Power

Produced	15,544,000 KW Hrs.
Purchased	49,288,000 KW Hrs.
Total Consumed	64,832,000 KW Hrs.

Reference 2

Chapter 17

ℰ✺ℭ

Garage

T he analysis was brilliant — his department was a service, a service that brought only costs and not revenue and would be the first to go — as departments were closed outsourcing the work.

My name is Larry Mitko and I live in Oakdale, PA. I was born forty-six years ago in Washington Hospital, Washington, PA.

I have twenty-seven going on twenty-eight years in the mill and was hired in 1969. Right now, I make probably . . . hourly rate is twelve-something an hour plus an average five dollars an hour bonus. Yeah, somewhere in that figure. I work in the Electrolytics Department in the Tin Mill.

However, most of my time was spent in the Garage Service. We provided truck transportation for anything that needed hauled between different departments. Also, we went to Pittsburgh, Cleveland, went down into the coal mines. We also provided the firetruck and ambulance when someone got injured, or if there was a fire in the plant.

Well, a typical day you would, early in the morning, you would start off by driving on a bus and hauling our trades and crafts, which would be our carpenters, pipefitters, you know, different people like that — laborers, painters — and we would haul them out through different departments all the way from the North Mill, BOF to the South Mills, Tin Mill. We would start doing that first thing.

The second thing once we got back — well, of course had our coffee — and then we would either get on flatbed trucks, dump trucks, and go to different departments — wherever you were assigned to for that day and haul whatever material that department would need to be hauled. So we were based in a central area, and then each department would request what they would need for that day, and then you would be assigned to do those duties.

Whenever an ambulance call would come in, whichever person was there, whether it was a truck driver or whether it was a mechanic. The first two people our dispatcher would see, he would grab and tell you that they needed an ambulance. Let's say at the BOF and that was your job, hurry up and get an ambulance going and transport the injured employee to our health center. When I went into the garage we had forty-four truck drivers, and I'm not quite sure, I believe there was twenty some mechanics, and that was back in 1969. We had, I would estimate, around thirty some trucks that serviced the mill.

Then you would leave the plant and go to other places. We had flatbeds that, if there was an outside truck driver going to come in for a shipment from, let's say, part from Pittsburgh and part from Aliquippa. A lot of times we would haul stuff either up to there to where the truck driver had to only go to one stop or vice versa, we would bring it back to here. We also had a parts truck for any time that we had any type of mechanical breakdown throughout the plant. His job would be to go and get parts needed to fix the broke down equipment.

We also had a medical car. A medical car was more of a service they provided to the injured employees, if they needed to come in for back treatment or any type of therapy or to see the doctor, or to see the nurse, or anything. We had a medical guy who worked steady daylight, eight to four, everyday. He would go to the person's house, pick him up, bring him in for treatment, and then take him home. A lot of times you ended up in Butler, Burgettstown, Little Washington, you went all over.

Well, in my opinion, no service ever pays for itself. All they are, are a cost, because you're not providing anything that you're going to make money off of. So in that aspect, it ran very smooth, and a lot of times if we didn't have enough people, they would just cancel a job and the department would order the truck for the next day. Yeah, as a matter of fact, they did bring outside contractors in from time to time. If all our dump trucks were broke down or all busy, then they would contract with either P.M. Moore Company or Unis Trucking. Unis would have a lot of dump trucks in there depending on the need of the plant.

Yeah, I remember a lot of good days. I think we got paid pretty well compared to other parts of the Beaver Valley, and just going in there and doing a good day's work was a good day.

Well, a bad day would be whenever you ended up with two or three ambulance calls during the course of the day. I would imagine myself, I probably picked up three or four employees, during the course when I worked in the garage, that died or even when I picked them up they were already dead. So there were a lot of bad days as far as that.

Right, I picked up people who had their limbs ripped off. You know, that was what I would consider a bad day, a bad day would not be a day that once you got there at eight in the morning you continued working until four. That's what you got paid for, so I wouldn't consider that a bad day. The bad days would be when you got on the ambulance or the fire truck and you ended up having duties that really would not . . . I'd say that you really weren't actually trained for.

When the mill was fully operating, I would say you would probably get . . . you would run into spans. You might go a week without an ambulance call, but then you might go a day where you get four. We were a 24-hour operation there, so a lot of times you would get these calls. Yeah, I think the Plant Health Center was very effective. Whenever I started in the garage we had four different, five different areas that we had nurses at. We had a main health center which was where we always took an injured victim with the ambulance, but I would say that it was operated rather smoothly. It was efficient.

So the health center was staffed around the clock. The main health center was staffed around the clock, and certain other ones years ago was staffed around the clock, like the Byproducts. We had a station over there and it wasn't the Byproducts, it was at our Chem Lab. Depending on the injury, automatically if you were on daylight and there were nurses and doctors over our health center, you would automatically go there. If it would happen on the night turn, depending on how bad it was, we had radios in our ambulances — we could talk to the nurse at the health center and explain what happened and everything, and how bad the employee was injured. We might take him directly to Aliquippa Hospital.

The emergency care was very good. It was better then than it is now. Because, first of all, J&L at that time could afford the luxury of having nurses around the clock and having doctors there on a regular basis during the daylight turns, because of the fact that you had 12,000 employees. Nowadays, we have a nurse that is staffed there Monday to Friday, eight to four. If we have an injury outside of that, depending on the type of injury, they'll either call an outside ambulance or they'll get a taxi to take him to Aliquippa Hospital. So in that aspect it has changed for the worse.

Well, probably the most important thing that I remember is that I enjoyed my job. You got to meet different people. You weren't at the same spot day after day after day. Other than that, as far as something to remember, I remember ambulance cases more so than anything, because that's something that you always remember.

Well, when the gentleman got his arm ripped off, he got it ripped off in an upsetter. I believe it was his last day of work, so something like that you never forget. That will always be in your mind, in fact, we just had a death last month in the Tin Mill. That brought back memories, I don't want to go into details, but the gentleman got hit by a train and got injured pretty severely, well, in fact he got killed.

I think that the company is really striving toward making it a safe place to work. I don't think that they wanted it not to be a safe place. The company and the union really worked together after the past contract, because they've introduced a new Safety Committee in which they have hourly people and salary people working toward a common goal of making it a safe workplace. So I think they have improved.

Well, I can remember a few fellows, because I still work with a few. Back in 1989 they closed the garage and we all got laid off, and it didn't look like we were ever going to have a job again. Some of them took the buy out and no longer have a job, at least with LTV. Other ones decided that they were going to wait it out and hopefully get called to the Tin Mill, which there are three other guys that I work with. So I remember those more so then anything. Plus, I remember being the assistant committeeman, and in the garage I can remember a guy got fired and he went to arbitration and got his job back, so that was another memorable event.

Well, normally I worked shifts, and then toward the latter portion, probably around '86, '87, I ended up with a steady daylight job on the parts truck. Then, probably '88, I ended up with a steady job on the medical car which was also steady daylight. The sequence of events here is again . . . but it seemed like in the late, say, 1977 there was large infusion of money and upgrade into the mill, probably 280 million dollars out on the island for a continuous strandcast facility. Then there were some combinations that took place, and now J&L was absorbed by LTV in 1974.

Yeah, in fact I went to Youngstown Sheet and Tube to haul lockers whenever LTV closed them down in 1976. All I know is that we all became one in 1978. It's like one unit. I believe it was Republic Steel and there were other plants that we started traveling back and forth to. We went to Cleveland, they had the east plant and a west plant, but it seemed to us at the time LTV still purchased them for the pension plan, but that's my own personal opinion. LTV purchased J&L for the pension plans. Well, that is my opinion, whether it's right or wrong, I have no idea. That was a fair assumption to a lot of the employees.

I noticed a change years ago, when I first started in the mill in 1969, and throughout the early '70's. It seemed that if a department was losing money, as long as the plant was making money as a whole, they never bothered that department. But it seemed like when LTV purchased them and became LTV Steel vs. J&L Steel, it seemed like if a department started losing money, then they would automatically shut that department down.

The first shut down began to occur on a large scale in 1981. Well, that's when they just announced that they were closing different areas. They closed the Blast Furnace, they shut the Blast Furnace down, they shut the BOF down, and it was just. I'm not aware of the exact years, 1981 was when it started. It just started like a chain reaction, our Rod and Wire was eliminated, the Byproducts was shut down (the Byproducts was one of the last ones that was shut down), but our Seamless was shutdown.

The last shutdown was 1989, and that was when they eliminated the General Maintenance completely. That's where I worked, it was considered part of General Maintenance and Service. 1989, that was the last portion of the shutdown. The last department, which was General Maintenance.

The Tin Mill stayed opened and has functioned ever since. The 14 Inch Mill, I believe, shut down for a very small time, but I'm not positive that they did. I know that they went, and LTV sold it to the people that own it now, as J&L Structural Steel. J&L Structural Steel is unrelated to LTV and J&L. The only relations that we have is that we still have the same union.

Well, the union had to downsize just as much as the company did, so that was probably the biggest change. You had to start learning that you had to work with the company. It's not like you could shut one department down and go to the next one. Well, now the Tin Mill is the only one there, so you had to work with the company or you are going to be out of a job forever.

Well, the old union, you could be more demanding, so to speak, because you had 12,000 people working on your side. Now that you have around 500 Tin Mill employees, you have to really work at both sides. You can't just stand and say, I'm the union, you're the company, this is the way it's going to be. You sit down and negotiate and try to make it run smooth. I think it has changed somewhat, but years ago the unions were really strong and more powerful. Well, at least in Beaver County I think you had more to fight with versus now.

My last day with the Maintenance Garage. Well, it was a sad day, of course. Everybody was really concerned. A lot of people, such as myself, didn't take the buy out, you know. I'm still working for the Tin Mill, but there were a lot of ones that did and it was a tough decision for everybody there.

The buyout was 1,000 dollars for every year that you had in the mill up to 25,000 thousand, so a lot of people had to make a decision. If you took the buyout, then you could not be recalled, or you didn't have recall rights. Whereas if you didn't take the buyout, then they would recall you based on seniority. At that time it didn't look like anybody was going to the Tin Mill. You needed like forty years to get into the Tin Mill Labor Gang, because of all the other departments that had shut down most people went there.

Well there was probably a fifty-fifty. Some of them took the buyout. If you were eligible for retirement, then you got your pension. But there was a lot like myself that, at that time, I had approximately twenty years, twenty-one years. All you can do is vest your pension, so there were a lot like myself. Like I said, there were four of us that went to the Tin Mill. Plus I think there were two mechanics that went also, but the rest either took a buyout, or took the buyout and had a pension coming.

Well, the early retirement, we had different plans. We had a Royal 65, we had seventy to eighty plan, it just depended on which one you qualified for. The Royal 65 was if your age plus the amount of years that you had in the mill added up to sixty-five then you were eligible for retirement with a kicker. The kicker would be like 400 dollars a month until you reach social security age. The seventy to eighty was just different years of service and age. I'm not quite sure of that because I didn't come close to any of those.

Well it was negotiated through union, not the company. I'm not sure that they would've just come and said okay, we're going to give you something. So it was the union that negotiated the compensation. Right, and it was a joint venture between the company and the union.

I would tell them to stay out of it. Only because I don't think there is a future there. Years ago, you used to have apprenticeship programs and you could really go in there and, if you wanted to make something of yourself, you could do that. Today, I'm speaking of the Aliquippa Works, all we have is the Tin Mill now. Maybe if you went to Cleveland, because Cleveland has Blast Furnaces, they have Byproducts, they have different departments up there. Maybe up there the possibility is there, but I say sooner or later they will downsize also. So I wouldn't recommend the mill to anyone.

Well, our steel is still going to be made here in the United States, in different areas of the country. Why sure, we're still strong in steel. Years ago they did a lot of foreign imports and I think, that really hurt the steel industry. Plus, we were overstaffed as far as salary personnel and hourly personnel. I think that the company and unions both negotiated and sat down, talked quite a bit about downsizing, and I think that it's worthwhile for both groups.

Well, I believe with LTV, once LTV purchased J&L, that they had too many plants and what happened was they started downsizing. Aliquippa just happened to be one of the ones they downsized on. Right now, the Cleveland Works is really working, they're booming. Yes, they downsized steel before all the other industries did, at least to my knowledge. Other companies are going through now what we did ten to fifteen years ago.

Byproducts dock

Management of barge dock

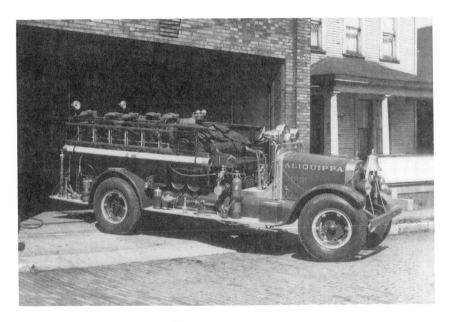

1938 American LeFrance Firetruck

1950 Seagrave Sedan cab

Chapter 18

ℰᎧᏩ

Labor

He was one of the last men I talked to, asking if there was anyone else . . . There seemed to be no more volunteers until he lurched from his chair in the back of the room. He was unsteady, as his wife tried to restrain him apologetically, don't . . . he shouldn't do that.

As he went on to tell his story, harsh and raspy in a whisper due to an old stroke. It wasn't a pleasant tale of a long lost love, but a painful remembrance of the work he hated. However, his sense of purpose and accomplishment was strong, he had survived.

The first time he had talked about the mill since he left ten years previous.

My name is Guido Colona and I live in Hopewell. I'm seventy years old and was born in Patra, Italy. My brother came in 1946, and I came after.

I was hired in 1948 and retired in 1983. I worked thirty-six years and was making a dollar ten cents an hour when I started, and I can't remember when I retired.

I worked in the General Labor Department. The General Labor Department repaired all the things everywhere at J&L. We went from anywhere in the mill and took care of any repairs. I get 'em out of fleet, I was truck driver, and took out the worker people to the job.

When I first started, I worked with a pick and shovel. I dig everything. So before we had machines, we would just dig up everything. We would dig up the Blast Furnace down in the pit.

Hard work it was. The crew was six guys going in every half hour and six come out. Take a rest and cool off or you couldn't work. Days were long — twelve, ten, eight hours, longer at the beginning.

Guido Colona

Sometimes, good days but everyday was rough. What could be happen? Nothing good, just hard work. So when we were in the furnace with pick and shovel and came out and our pants were on fire, and our shoes were on fire, that's bad. It was dangerous. There was fire.

I was driver, and liked those days when I took the workers to the job. Tools, people, everything, we took.

I would like to forget everything. I don't want to remember nothin', all those rough years, rough, forget everything. It was rough for me. John Bisotti was my foreman, I liked him. John Struthers was the boss, and George Abbott was senior.

Rough day everyday, five to six days a week, no easy days. There is always somebody loafing, but not many. I liked working for John Bisotti, yes, George Struthers, no. He didn't like the foreman. He always watch

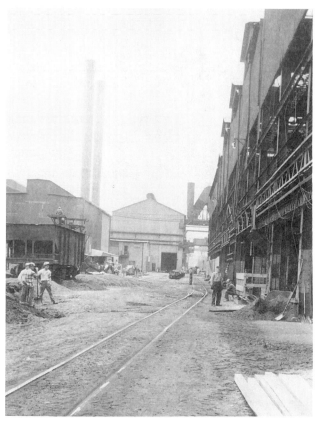

Pick and shovel work

you, what can you do, you just do your work and that's it. I like my work, but not my boss.

Sad, I'll tell you, one day I was under the pipe on time, maybe 1950, when I started. Yeah, so much pipe fall you get scared.

More nights they had to call me, they say get out of bed you have to come in.

I had a son that died at nineteen. I had a wife, divorced in 1945, and a kid in Aliquippa. I went back to the old country, and then I got married again. We lived in Italy for two years. My first wife, she's around. Okay, and my son died after, he handicapped. He was sick and then he died.

I would change everything, keep it open, that's it. There's no more. It was all right to keep it opened. My last day I was driving and they called me up to me to retire early for extra money. So I took a pension.

I can't talk about the mill anymore. Don't talk about it at all, not since 1983. Yeah. I'll talk over here, first time to you, that's it. Too bad down there.

Rolling Mill Labor

Labor Hi-lift clean up

Slag alley

Furnace rehabilitation

Foundation excavation

Clean up detail

Chapter 19

ဢဢ

Machine Shop

H e was the "organizer", the cohesive component of the group that
kept those old timers together like it was 1937 again. He motivated
them when they were uncertain, bringing rules of procedure to their
meetings, discussing political agendas or what health insurance was best
for old agers. As a youth he played baseball and semiprofessional football
as an Aliquippa Indian and proffered his fifty year reunion photograph.
In his later years, he took his group of old timers to plant shutdowns and
contract negotiations so that the lessons of the past are remembered and
not forgotten.

My name is Joe Perriello. I am eighty-one and was born in Saberton,
West Virginia. A small town outside of Morgantown.

I worked from 1933 to 1977, so what's that, forty-four years and
somewhat months. July 1933 and retired in September 1977. I was
making thirty cents an hour in 1933 and when I retired in 1977, was
making six dollars and ninety cents I think. I worked in the South Mill
Machine Shop. Well we did general repairs for the Tin Mill and the
Wire Mill. If the other departments needed us, we helped them, too, but
it was primarily the Tin Mill and the Wire Mill repairs. The South Mill
Machine Shop differed from the North Mill Machine Shop. There was a

Joseph Periello

different type of work up there. Their stuff was mostly Rolling Mills and our stuff was Finishing Mills. The Tin Mill was finishing, the Wire Mill was finishing, and the Nail Mill we had for years, but they did away with that. I think they had a little heavier things to do up there. They had bigger machines and heavier stuff. We did a little different kind of work.

I was regular machinist standard down there. Then I was the shop steward, four years. It took four years of apprenticeship. I started in 1937 and I finished in 1941. We started and I was working as an oiler at fifty-six cents an hour. I think, but to be an apprentice I had to start at thirty-seven cents an hour. So I lost money. Then every six months or 1,080 hours we would get a five cent raise. Five cents every 1,080 hours. Top rate was sixty-nine cents an hour.

The best day in the mill was whenever I was instrumental getting the wages for machinists even. I was instrumental in getting George Fry, our industrial engineer, Bronko Maravich, and Tony Vladovich to get the company to see what was happening down there. The machinists were getting sixty-nine cents an hour but there was a turn lathe man making a dollar-fifty an hour, because he was an Elk and the guy down there was an Elk. You know, they were buddy buddy. In our department there was forty-five or fifty machinists and there was ten different rates. It was impossible to get a raise. I worked eighteen years as a . . . what they call . . . I wasn't a standard, I was a beginner. Then the intermediate rate, then the standard rate, see, and I worked eighteen years at sixty-nine cents an hour. There is no C, B and E.

Finally, I argued and fought for it. Then, in 1956, I think it was, they finally agreed to standardize skilled labor. They started us at twelve, fourteen, sixteen points. See, when we became machinists we got to be twelve points then whenever you got the intermediate rate it was fourteen and then standard you were sixteen. We were all paid the same.

It didn't matter if you were an Elk, or if you were black or white or Italian or Serbian. That was in 1956 when they came out with that. It took a long time. It took a long time, that is the thing I really was proud of. We were . . . we had a pretty close knit group. We worked our eight hours and we took our half hour for lunch, even though we weren't entitled to it. But we worked what we had to and whatever we didn't have to we didn't. I never liked the cost of living raise. I always argued the point that it was built inflation and it was. I didn't like thirteen weeks.

The thirteen week vacation in the steel mill, that was diversified, it was almost impossible. Because they didn't give you the thirteen weeks like department by department. They gave it to you throughout the whole plant. I might be entitled to thirteen weeks in the Machine Shop. Say, they allow a hundred of us to go that thirteen weeks, well, ninety-nine of them might be throughout the plant. There might be five or six in the North Mills, five or six in Tin Mill, five or six in Wire Mill. If you had an individual job it was all right. Like us machinists, we walked out. What the hell, that was all right. Of course, if you're working a crew and if they took a man of that crew, then they put a greenhorn in there, the efficiency of the production went down.

I don't have the faintest idea why they did it. The company and union finally realized that it was wrong. You had to take the thirteen weeks when it first came out. Then they tried to give it to you. They started shutting the departments down for two or three weeks a year and count that toward that thirteen weeks, and they tried all kind of things but it just didn't work.

My experience was good, it wasn't bad in our shop. Fifteen minutes. We were allowed fifteen minutes, and it wasn't that bad, for cleaning the machine and go upstairs and change clothes, wash-up and change clothes.

They wanted you to stay the hell away from the clock, don't line up ten minutes before. If you get done, sometimes you sneak up there fifteen to twenty minutes early. They didn't want us down at the clock, even five minutes before waiting around.

When I found that one of our supposed-to-be friends was a stool-pigeon. I worked the Nail Mill coiler. We had high priority on brass and stuff during the war, this was 1941 or 1942. In 1944, that's when I turned in my time and went out. He would send me up to dump the brass in front of the boiler room, where they kept supplies. They dump maybe tons of it. Three times as big as this room. But he send me up and he say, Joe, we will sort it out, we're going to have to sort it out by pattern. You go up and it take like three or four days. You'd have a wheelbarrow and you take it from the pile and select them all, and by the time you get . . . When there was a big pile it was easy, but when you get down to a small, you have to go looking for them, and it takes two or three days. I understand my boss was looking for me.

See, I worked for the Nail Mill coiler. My boss was a machine shop boss, and the guy that was supposed to be a friend. We did everything, we used to play softball together, we bowled together, and all that stuff. Then finally, this one day I came back from putting away the stuff and my boss said, where the hell have you been? I said, what do you mean where have I been? He said, I asked Red and Red didn't know where you were. He said, you were screwing around somewhere. He said, you have been gone three days. I said, look Jack — his name was Jack — I said, I was down there sorting out the brass and Red knew that. And he looked at me for a minute and then he said, you know, you're a dumb f'in dago.

He said, he has been stabbing you in the back for years. Everytime you quit a little early, he said, he tells on you, everytime you leave he would talk about you. We didn't have a clock, we had a check-number. I worked to, say, three thirty, regular turn started at eight. They worked eight to four, I worked from seven to three. So they would come in and open the check box and take my check, and maybe sneak out ten minutes before or ten minutes to three. He said, don't you know, he has been stabbing you for years. I said, why. He said, because he's that kind of guy.

So I went over to see Red and I told him, what's the idea, why did you do that to me? Well, he said, when it comes between my job and your job, it's going to be you. I said F- you, that's what I said. So I was having a little problem at home, you know, marriage crap and all that. I was ready to give up anyhow, so I put in my ten day notice and I quit. I went into the service. Well I was gone two years and according to the GI Bill of Rights, it took me three months to get into the service. What happened was that I didn't have much luck, talk about worst days.

When I went out to join, I wanted to go into the Navy. They told me I couldn't go because I gave my ten day notice, that threw me into like the first of April. The first of April they changed the rules. You couldn't voluntarily go in, you had to be drafted. So I had to go and get a job, so they hired me at the Democratic Club for three months, until I got drafted, and I wrote a letter to the draft board before I even quit. I said, you know, I'm not married anymore, I'm this and that, and I think that being single, I should be drafted. So finally the draft orders came and I went out.

When I came back, they didn't want to hire me with my eleven years seniority, they said I quit. I said, I didn't quit according to that Bill of Rights, I joined the Service. They said, yeah, but there was three months there that you didn't work here. You went to work somewhere else before you got drafted. I said that's the idea, I said, I got drafted because they changed the rules on me. So, I told them I can't lose this eleven years seniority, so I left, jobs were pretty easy to get in those days. It wasn't hard to get a machinist job.

I went home and told my mother and dad, I said, you know they wouldn't hire me. I had moved back home after the war. And they said, well, what are you going to do now. I said, I have two or three friends that said that they could get me jobs elsewhere. I could go down Westinghouse. Bill Pastine, he was the president, he and I played football. When I got home, they called, J&L, they called back down and they said we'll give you your seniority back. They must've called the draft board and they must of read that letter that I wrote, and they gave me full seniority back in the shop and in the plant, and that was all right, but the company never challenged me anymore.

The higher up I got in the union, the more the men challenged me. Because they resented the fact that I was in the union, that I was gone for two years and came back and got full seniority and all that. They insisted that their seniority was greater than mine, because I broke mine by quitting, you know. I had a hard time down there, there was a couple of guys, as far as the union is concerned . . . The two people that gave me most trouble were the ones that were conservative folks.

Say I started my apprenticeship in 1940, right, these two guys started theirs in 1941. Well, Chuck and Johnny started in 1940 but they got drafted. They served four years in the service. When they come back, they got their four years of apprenticeship. So now it came to the argument who was going to stay on daylight. I said they have to stay on daylight and they said, oh, we were machinists first. I said you weren't according to the GI Bill of Rights. If they would've stayed here they would of been machinists one year ahead of me, so, of course, they wanted to hang me, you know. We had that problem. Every time you were . . . well, if you got a soft job — and you do get those once in awhile.

I spent time in the office, because what the heck, that's where you did your work, you know. Talking to you, didn't do me any good. I have to talk to the boss there to get something straightened out, right. But we did a lot of things down there that . . . just like when they got the readjustment agreement. I got my grinders and my drill press operators two points more than anybody else in the plant, and everybody wondered why. I didn't sign the agreement until they came through, because I said they were actually just behind machinists. Why do machinists get twelve points and other operators did the same thing?

The relationship between the union and the company in the early days wasn't any good. In the early days, like in 1937. You see, all we won in 1937 was the right to vote. So we got the right to vote and be represented, and from 1937 to about, I would say 1950, things went slowly. There were all kind of arguments, you couldn't win an argument no how. In fact, what pissed me off more than anything else was my first five years. I was hired July 19, 1933, so when vacation came you get one week for five years. So they cut me off because they said you had to be hired before July 1st. So in nineteen days I lose that, and I had to wait six years for my one week. I came back from the service, now I'm entitled to two weeks in 1946. I couldn't get anything, those two people — Mike Brummitt was president and this Chuck Davies he was secretary. They signed an agreement where if you weren't back by July 1st, you didn't get any vacation. So I got screwed again. So I had to wait twelve years to get my two weeks. So I got pee'd off.

Then we had a strike in 1948 and on the picket line with me was Bronko Maravich. Bronko was a committeeman for Seamless Tube Mill at the time, and I was on the picket line. He was the captain of the picket and they were letting people in. I said, Bronko, what are you doing? Well, he said, these people are general maintenance. We have an agreement that they are going to do certain work. This strike was only for pensions, and I told him, how could we have to keep these guys out that have three or four kids, and have these single people go in. Well, that's not striking. I said, I used to picket line to keep the one guy out that I knew has two or three kids, but somebody else would go in there that is single. Most of these guys were single that were going in there. They were all carpenters and bricklayers, pipefitters and stuff, you know, so I left it wasn't a fair picket line.

So then when they asked me join the Clean Sweep, I joined the Clean Sweep, and then, I think . . . that was the gang that swept out the Davies and Brummitt out there and I think from 1950 to 1958, while we were in power, we had the best union representation ever. Tony Vladovich, Dave Johns was financial secretary, Bronko Maravich was recording secretary, and we got more done, more respect from the company. Even I as an assistant committeeman had as much of a say as anybody. That's why I think it was the most beneficial as far as the union is concerned.

I would like to forget Red in 1946, I would like to forget that one. He was turning me all the time and me not knowing. I thought we were friends. You know, you watch out for them. Like a guy told me one time, he said, it's not your enemies you have to watch, it's your friends you got to watch. You expect it from your enemies, but your friends.

We had a problem with the night-turn boss all the time, and he kept harassing certain people. George Lane, at that time, was the master mechanic of the South-End. This one time I was scared and I had a problem with Bill Zinz and I kept trying to say, look, we have to do something about this man. He is picking and always griping and he started at five o'clock. Let's say he put you on the turn lathe or on the Boring Mill or somewhere. He would come in and take you off the job and put you somewhere else. So now, when these boys that had a problem with the apprenticeship . . . You know. His name was Zwigart, he couldn't get a B-rating — this is before then.

The boss said, Johnny, you can't run a B-Lathe, all you can run is the Lathe, you got to run the horizontal Bar Mill. So I said, well, put me on. Bill don't start 'til five, John starts at four, he is on night-turn. So Shivers is the boss until five or five-thirty. So he puts him on the Boring Mill so he could get a B-rating. So Bill comes out at five-thirty and takes him off and John doesn't want to get off the machine. So he calls the cops, they drag him out of the mill. They were worried they were going to draft him if he lost his job, he got two kids. So I come down and I say, what the hell is going on here. So I said, I'll tell you what, I put my tool box on the corner of the bench, nobody works tonight. All you guys sit on your ass, he said, what's going on. I said, nobody is working tonight 'til we get this shit straightened out. They going to railroad Johnny and he only did what you told him to do.

So about five o'clock George Lane come down. He said, Joe, I understand I'm not going to have no night-turn work. I said, that's right, I said, I got my toolbox here. I said, I know, it's the position that I get myself in and you can fire me for doing what I'm not allowed to. In my position, I'm not allowed to call a sit-down strike or strike, it's right here. But I'm telling you now, this man is being used.

I explained it all to him and he said, put your toolbox away. He said, it will be all right. He said, what do you want, and I said, I'm going to tell you what I want. I said, I want Bill Zinz to go on daylight for a month and put Carl Brunner on night-turn. Let's see what the hell goes on here, let's see if these kids on night-turn will produce and will work without this damn harassment day after day after day. He said, okay.

The result was they put him on daylight, and on night-turn things got better. So now, he is sixty-five years old, he supposed to retire. So I tell George Stet he was supposed to retire. He said what do you mean, he is going to retire. I said, hell no, he bought a new pair shoes, and that tight bastard wouldn't spend a dollar if it would kill him. He would walk to work from Ambridge, he lived in Ambridge, he walked to work from these.

We worked in the South, we knew where the bridge went, and that was where the shop was, and he would walk from here to there. If you had to contribute a quarter to death benefits, he had to wait until the next day, he wouldn't bring it. I said he spent thirty-five dollars for a brand pair of shoes. You mean to tell me that he is going to retire, he's not going to retire. We had a pretty good rapport, you know, we talked, the guys.

In the mill, it was the guys that got me started in the union. It was the Colangelo boys and John Nellish and John Babich. Then the man that I really . . . When I first started, see, I was with the old amalgamated in 1934, 1933 which was what iron and steel

There was the Amalgamated Association of Iron, Steel and Tin Workers. Yeah, when we started that and another union, the Steelworkers Organizing Committee, they merged sixty years ago, too. I was scared shitless. My mother, too. We played baseball with members of Iron and

Steel and I was on daylight. I started working in 1933 and we started a baseball team as Amalgamated in the mill in 1934 and 1935. The unions were trying to organize. So they wouldn't let us play in Aliquippa. They run us out of town, to South Heights. We played maybe ten or fifteen games.

Now, like I said, Phil Pastine was president of Aliquippa Indiana Club. I played football for the Aliquippa Indians and from 1933 to 1937 or 1938. In fact, those pictures I gave, you that one picture was at the Indian Banquet. See, I came home at eight and went to bed about nine and some cops come over and threw me out of bed and wanted to blackjack me. It was Phil Pastine, and he stopped and he said, leave him alone, he stopped talking.

Because I was playing with Amalgamated Union baseball. They didn't want me. So these were two company cops and one Aliquippa cop. Phil Pastine was an Aliquippa cop the other two were company cops. They had that kind of power in the 30's.

Oh yeah. Jesus Christ, they had power to do anything they wanted. If you got arrested for drunkenness or anything like that. They take you to jail, take you to the Justice of the Peace, you sign a paper, and they took the fine out of your pay. You never had to pay them directly, they took it out of your pay. They did a lot of things, you know. You talk about this OJ Simpson trial, talking about Rodney King, we had that everyplace in Aliquippa — white people get beat up. I saw people get beat up, their head split open and everything else, and a cop walking right down the middle of the street and not put him in jail.

It changed after 1949. Because the union come in and we started both voting Democrat, see, you couldn't beat 'em. When I was . . . well, another thing that happened, now, I'm nineteen years old and I'm hired, right, 1933. The November election came up, I'm nineteen years old, I'm oiling on the old Hot Mill. So they come over and tell me, Joe, get in the car, you're going up to vote. I said I can't vote, I'm only nineteen years old. Oh no, you get in the car and go vote. I was supposed to go to Logstown, Logstown School is Precinct #1, where I lived. They take me to Franklin Street, that's Precinct #3, I go in, and there I am, registered and voting. Then the next primary, and the next general in 1934.

One of the drivers that they designate take me to vote, you know. Most of the time it was the union representative from the company. The company union person. Yeah, but the second time nobody took me to

Logstown, they took me to my own precinct. That's guts they had there. I was there and registered as a Republican. Well, sure I would've voted Democrat on my own both times instead of Republican. So then that changed once you had the union.

That changed, that changed after 1937. People don't understand what we did in 1937. When they gave the right to vote for a union. We had to beat the majority of all the people that worked for J&L. Everybody, that was all the supervisors, all the board of directors, everybody that didn't come down to vote was a no vote, understand what I mean. If you didn't come down to vote it was a no vote and we still got eighty-five percent of the votes. We got eighty-five percent of the votes to get the union. To get a union isn't that amazing. As a no vote, we had a hell of a lot of people that weren't rank and filers. Well, my family that was another problem I had with Salerno, the company. My father and mother were born in Italy. My father . . . this is amazing, you know, the company used to always talk bad about the foreigners and the courage that they had leaving their mother country without being able to speak or write the language. Well, my dad left Italy when he was about eighteen or nineteen and went to Brazil. He worked there five or six years, and then he went back and married my mother. They knew each other when they were kids. They went back to Brazil and had two sons, my older two brothers. Then they heard about America, so they came to America. Now this is amazing, is that something?

There are two people who couldn't read or write. My mother was illiterate, my father went about third or fourth grade and they came to America. Now, I don't know whether they stopped off in Clarksburg, West Virginia, but they ended up in Saberton, West Virginia. That was where I was born, and my brother, Chuck, and my sister, Mary, I think, unless they were born in Charlestown. They had a Tin Mill there and most of my family settled in Saberton, New Kensington, Wilmerding, Aliquippa, McKeesport, and New Castle, and my dad had heard about Woodlawn, the old name for Aliquippa. So he came to Woodlawn. I don't know whether I told you or not, but did you know how people were treated before the union?

Well, I'm going to tell you. Now these two guys, they are real steel people, Jones and Laughlin. They bought all the land from Monaca to West Aliquippa, skipped Aliquippa, and then all the way down to South Heights. Surrounding areas like a horseshoe, they took all the surrounding area from South Heights to Hopewell and Center Township. They built the steel mill, and they asked if he could come and work. So people said that my dad came. I remember 1919, flu epidemic, I was five years old. We were in Woodlawn then. They hired you, then and they gave you a job. Depending on what nationality you were and what color you were, that's when they separated you in different plans. Our people ended up in Logstown, Plan 11, Plan 7. West Aliquippa was a town by itself. They had company houses.

Plan 6, Highland Avenue, Franklin Avenue, was bosses. They were all cake eaters, not bosses, cake eaters. Plan 6 was the bosses, that was the big house. Plan 12, Plan 8 was all cake eaters. Then they started, well, here is a house, we'll give you a job, you live here, you don't have to worry about the rent. You don't have to worry, we own the house, don't worry about the water, we got our own water company, don't worry about eating, you go to the company store and buy what you need to feed your people. Need clothes, buy them at the company store, need tools, buy them at the company store, don't worry about nothing.

Just sign this paper and we can take it out of your pay. You know, there were maybe thousands of people down there who never drew a dime of pay cash until 1950. Three times all their working life, do you believe that? You better believe it. This is the way they got their money. They got a few dollars cash. They go to the company store and say they get ten dollars worth of food and they would sell it to you for seven dollars and fifty cents if they were lucky, that's the most they would get. If they were desperate, they give to you for five dollars. That's the way they got a few bucks back. That's where they made their dimes.

So basically, then, with the company store, you took all your pay in scrip. Everything. They gave you the house, they gave you the water, they paid the gas, they paid the electric. Well, there was no electric, you paid your transportation. You could eat in the Mill Restaurants, you could buy books, and all that and the cops would meet you at the train station. You could walk to the station and they ask who are you. You say, well, I'm Joe Perriello. He said, oh, where do you work, in the Tin Mill, okay. Who are you? Mike Keller. Where do you work? I don't work no place. I'm with John Brown, yeah, whose your friends. Well you name this guy, this guy, and they say you MF'ker get on that train and get the hell out of here. You had to have somebody vouch for you as amalgamated.

My father came in on a . . . imagine what my mother and three bothers and his brother were all the brother-in-laws. Then he brought his brothers over and piasan. The land Callebnia, I think there were twenty different families he brought over here. They had to have some place to come to or they couldn't leave the country. So my dad always had one or two people come in from Italy, living in our house, and never charged them a dime till they got a job somewhere, you know. Until the day he died, he had all those people. They would come to see him two or three times a year.

But he had bad luck. Two of his sons got real sick. My brother, Nick, was a great athlete. My other brother, Dominique, got spinal meningitis and he got to the point where we had to put him away. So he lost two of his boys. He was a gutsy guy, he stayed right in there. Yeah, he worked the Tin House. Now, good thing he was working twelve to eight. He runs a tinning machine. That is the last thing they do before they sort. All the tin they push and go through, you know, this was steel plate going through tin, and then it goes up through a roll and it gets it all cleaned up. The thickness is judged by how much and how long it goes in there or whatever.

Some might be 1,000, some might be, it all depends what the tin was for. Well, this one day, a piece of tin curled up and cut my father between and tore two fingers, right past the bone. A good thing he was working twelve to eight because he had to go to Southside. So the doctor up there patched him up the best he could and sent him up to Southside, and then they sewed it together, and his fingers were crooked like this for the rest of his life. Never got a dime for it, and he said they said he could still work.

You know the worst part of the steel mill is changing the turns, working night turns. If I had . . . if I was a super-being, I would have everybody working daylight. Night turn caused a lot of problems. The wives, you know, work four to midnight for five or six years and there is guy that work twelve to eight for four to five years . . . I'll tell you, there are so many divorces caused by that, because the guys, they actually got sick physically from working shifts because they were worried.

Say what we told the young people then, get an education. This education didn't just start yesterday and the day any of us walked in that plant, I told my son, my father told me, my brothers told me, we told everybody else — go to school. Get an education. Don't be a mill hunky like me. See what I have to do, look at my schedule, I have to work daylight, I work night turn, I have work Saturdays, I have to work Sundays, Christmas, Christmas Eve, Christmas Day, I'm working. Don't be a mill hunky like me, go to school. That's the best thing that came out of World War II, you know.

The education, Christ, we went to Penn State one week out of the year. There we talk about the new, talk about the past, talk about the future, you know. There was 10,000 students from trailers, because they didn't have any room for them in their dorms and fraternities. This was Penn State alone. Now we knew then there was going to be two kind of workers — high-tech and laborer — there could be nothing in between. What do you have? Those two things, right.

Well, I am very disappointed in the union. I thought that when we signed the "no strike" contract they pulled our teeth and the company knew that they could do anything, and we had to keep working. They didn't have to sign contracts before the steel mills continued to work. They sucked us in.

So by giving up the strike clause, we didn't have no zip, no power, no nothing, and I think they say Aliquippa was old. It was one of the smartest mills in the country. You realize, we had a continuous caster, Tin Mill, my God, we made tin by the mile. We made tubes so fast down there they couldn't keep up. In the Wire Mill they use to take 300 pounds and put on their heads and throw it from block to block. Now they got a seven hole mechanized electrically controlled wire, and all they do is stand on one end and weld piece after piece and kept running it through. We were blocking out 6,000 pounds of dead block when I quit, when we left.

When they first started the Wire Mill, they were making bundles by 150 to 300 pounds. They had to go from one hole to the next hole, so they had to pick it up by hand. You know, like I always tell the story, they were six feet tall when they started, and five foot six inches when they quit. Then they put on the buggies and then they came up with these rope machines. They had seven blocks, electronically controlled, all you did was weld on one end and you fed it through, see, and it would spin out. When they first started they were only taking off 300 pounds. When I quit, we had a 6,000 pound dead block. They could make 6,000 pounds of wire without changing threads.

The story that they tell is that back during the oil crisis the Seamless Tube was supposed to be done away with. They couldn't maintain their share of the market. When the oil prices came up, the goddarn Seamless

Tube became high priority. All the steels they made, right down the Seamless Tube. There was nothing for the Wire Mill. So they shut the Wire Mill down. If they would've kept it running another six months, it would've never shut down.

After that, I think they used very bad judgment, they had superior people there, they shut down the best steel mill in the world. I still think it was because of having a strong union made that plant. Well, my thoughts are that a lot of good things came out of Aliquippa due to the union and the Democratic Party. I think that it turned out that the Democrats involved in government were just politicians. I think what happened from state to state, they outbid each other. I think that is what happen. I think Cleveland somehow, rather Ohio, outbid us. For the business and LTV. LTV went down the drain.

Oh, boy, when they went bankrupt that was terrifying, you talk about a black day in my life. When I got up Monday morning and they said no pension, no health benefits, no nothing. The response was that was when I finally found the worth of the union. That's when Tony Rinaldi went to Hennepin Wednesday that week and he got them to have a strike vote. That's when Blackie DeSena called me about ten a.m. Monday morning. He already got the golden dome at Beaver for a mass meeting on a Sunday. We had to go down to the Union Hall and Rich Vallecorsa gave us a hundred and fifty dollars for the insurance. We had to have that to pay for the . . .. But by the time we got up there on a Sunday, Tony Rinaldi already had Hennepin on a strike vote, and LTV reconsidered and gave us back our health benefits, you see.

The government gave us our pension, they cut the hell out of it, though. Unions efforts and SOAR and us old people, we did a lot of traveling. I was on the bus so much I got a sore ass, I'll tell you. We went to New York three or four times, went to Washington three, four times, we went to Cleveland, went to everyplace, eleven buses at one time. The first time we went to New York, we had eleven buses.

Young Joe Periello Labor Day drawing

He was a jolly man, round features, ever-present smile and laugh, as he gleefully wove a tail of immigrant success and advancement in the post-war American economy.

My name is Chris Trotta and I live in Aliquippa. I am seventy-two years of age and was born in Italy. I started working in the mill in 1954, in November, and worked for thirty-three and a half years. My hourly wage was eighty-nine cents, I think, an hour when I started. I retired in 1988 and we had wage cuts, we were making around ten dollars an hour.

I started out in the Welded Tube department and I moved around in the mill. I ended up in the Machine Shop. Yeah, general maintenance. The Machine Shop, they make parts for support of the mill in Aliquippa, and they made products for Cleveland and products for Pittsburgh. So basically this machine shop manufactured the parts for the production machinery. I started out as machinist helper and I was working as a machinist helper, running the turning lathe #12, and some small parts manufacturing.

Right, and then when I retired I was around the cranes. They eliminated some jobs and combined some other jobs, and they gave us raise, so they could move us from one job to another. You worked crane and if you weren't needed you came down and worked somewhere else. It

Chris Trotta

wasn't very often that they did that. Saving money, naturally, was the angle behind that, which was better for the company. They gave you one raise, then you had to go where ever they told you.

Well, the best experience . . . if everything went smooth sometimes it really went smooth, sometimes we would go with machinists to different jobs in the mill. Sometimes it goes pretty good, sometimes it went pretty rough, it all depends. The best day was when you went to the mill, put your eight hours in and then went home with no trouble, that was the best day.

To become a machinist it took three or four years but I did not go. At that time, at my age, it didn't pay. I would've had to put all the rest of my days into night turn and I couldn't see that.

Well, the worst day was when you went to work and everything isn't goin' too good and you got hell from the foreman, usually as others have, I'm sure, and that made it even worse. They were the worst days. Well, the foreman had responsibility like anybody else. So if one thing go right for us sometime it don't go right for them, but if one thing go right for us they saw. But when things go right for them, we didn't see them. So they probably had a superior, who did something to them and was passed onto us.

In reality, everybody had responsibility once you walked in the mill. You had a job to do, if everything went smooth, you did your job and nobody bothered you. If things didn't go right, well, then you had a problem, and sometimes my problem was things not goin' right or something wrong or maybe the machine that day, maybe you had bad steel. You could've had a variety of things that could go wrong, and if he wanted something he wanted them now — not next week — and if you couldn't get things goin', like anything else, then it was like you typed a piece of paper but had to throw a half a dozen away. It happened.

Well, I remember one day when they had brand new mill in the Welded Tube, they were testing pipes on the Electric Weld (EW) Mill. The tester was set off in the bay and there were guys watching. They bring the pipe from another mill and one pipe let go, had busted, and was full of water and knocked them guys off from the wall, knocked them down, and you see guys come down. You see a guy . . . that's very sad how you look at it, it's a sad thing to see when somebody tried to earn a living and then to get hurt sometimes. It had to be the fault of somebody, but the company always had policy that accidents don't happen. You make it happen and that wasn't always so. Well, there were accidents, I have witnessed a couple of them, two or three of them, but I'm sure in the mill, all over the mill, I'm sure it happened a few times during the day.

Well, I wish that we didn't have to try to forget the mill has been destroyed, because of how much I put in it. Somehow I wish the mill was still here regardless how bad it was. All of us made a good living off that mill, and I wish that was here for the future generations. It's sad, that was the saddest day really when they started tearing the mill down. I was sad. Somebody blamed politics, and in many ways they spend a lot of money in Aliquippa. That is my personal judgment, when they bought the mill from Pittsburgh they had a 64, 54 Inch Mill. They went and brought that one from Pittsburgh over here. That was no good if they would've had the big mill, I think this place would still be working. They spend all this money in the Steelworks on the island and there ain't nothin' there. Just, when you go by there it makes you want to cry.

There is nothing there, they just they put a highway over it they put a three lane highway — for what, for who? And after all that I still say the politicians are crooked — maybe this shouldn't be on the record. It's funny, they got a guy to sell all that property for 300,000 dollars or so, a figure somewhere around there. Then they come around and do the taxes for the company for over a million dollars, then they give him a break, then Aliquippa gives two million more a break, then they go down and give him some more breaks. Old American steelworkers or taxpayers paid their tax, and that guy is putting a mall in there for benefit to who? Sure the hell you don't want to help the working people.

There was a guy there name of Masgow. He was working one night, a piece of metal fell against his leg, he was taken to Aliquippa Hospital, and they and somebody did a bum job and when it was to late they took him to Allegheny General. But he died because of negligence of the doctor and I don't care who is, anybody can make a mistake.

Funny, I don't know if that would go over there or not. The funniest story that happened, it's hard to forget. We had . . . now you are not allowed to use the color word. But this lady was, unfortunately, it happened to be a colored lady, she was cleaning a machine. Well, she had a heavy jersey, she had nothing else underneath, and the machine was running and the chips were breaking away from hot steel and went inside her shirt and burned her. She started goin' like this and they are both hanging out. That was funny, you could never forget that. I know, but you'll have to change it around somehow.

No, unfortunately I have no family. I was married and my wife passed away in 1986, and never remarried. My wife, well, she didn't have to go in the mill. I work steady daylight. I could've earned more in a higher paying job because of night turn. I didn't like, so tried to stay on jobs I could work steady daylight. But I didn't mind four to midnight but I hated midnight to eight. I was drunk without drinking, I wasn't hungry when I was supposed to be, when I was suppose to be sleeping I was eating, when I was supposed to be eating I was sleeping. It was miserable but I make a living and that. I'm sure a lot of other people weren't happy with it either, but what are you goin' to do? That's the life.

Well it's too late to change anything because the mill isn't here anymore but I think a lot of things in the mill were bad for working reasons. When politics got involved in industry, one day forced the people to have different people, they weren't qualified. I don't care who it is, they don't deserve the job, but that's what I believe. Everybody deserve the same right, give 'em a chance. If he doesn't pick up the job, if he doesn't learn his job in a certain amount of time, I don't think the company should not be stuck, no matter who it is, should not be stuck with it.

The governments were subsidizing while they were training after the company was making the money on it, and then afterwards the govern-

Safety glasses to prevent grinding injury

ment said that we ain't paying no more so things started changing around. The last day in the mill I went to work in the morning, I worked on a crane, we quit at three. Around one thirty my boss calls me and said come off the crane and talk to your buddies and say hello to them and good bye. I was happy about getting out of there as I walked in the mill. I walked in the mill in fairly good health, and walked out of there the same way. I was happy for that, and then I was sad that you spend a good part of your life in there, and all a sudden you're not a part of it no more. It kind of saddens you, I don't know, it sadden me. But what I miss more are the people I worked with, they were a nice bunch of people. There were some sour grapes in there, but no matter where you go you find sour grapes, but you expect that in any field of work.

Young people today, they are coming up now I don't know why they're lazy. They might not be lazy but a lot of them have no ambition, they have no lives. Some of them are great but a lot of them just don't have the drive or anything to go look for work.

Well, let me put it this way, I was once told by an older guy than me that a chicken should always feed what it scratched and if you want to eat, you have nowhere or nobody to feed you, you got to find it right.

Leni Vukmir

But because there is somebody that supports you, that makes them probably not lazy, but makes them, I don't know, they have no drive.

This is still the greatest country in the world to my estimation. I have seen a few countries. I was in North Africa, South Africa, Pretoria, Kenya. I was in England, I went through France, Switzerland, part of Yugoslavia, was in Germany, and this is the best country in the world that I know of. If anybody has any ambitions, you want to work, you can still make it today. It's not like yesterday, but still today in America when they say God Bless America. They better pray to the Good Lord and say God Bless America everyday, because if something ever happened to this country these people would never make it anywhere else in the world. What else can I say.

She was fiercely proud of her independence and ability to return to the workforce after the loss of her husband in order to keep her family together.

My name is Leni Vukmir and I'm sixty four, going to be sixty-five. I was born in Steubenville, Ohio. I had six or eight years of service and started in 1975 or '76.

I was laid off in 1981. The most I made was 18,000 dollars in one year. Well, when I started, I didn't start until January. I worked in Labor in the North Mill in the Machine Shop. I cleaned and shoveled chips. I put it in the wheelbarrow and hauled it out and dumped it. Come back, did this all day long.

Basically, I had to clean up the chips that were around the machines and put them in the barrel. That's what I did all day long, and if the chips were cleaned up a little bit, then I had to sweep around the place. Very much hard work. Basically, with most of it shoveling all those chips out. You saw me. Remember when you were in there? You saw me, you were so upset. You had tears in your eyes when you saw me and I told you don't worry about it. I remember. So basically, most of my work was cleaning up the chips and putting them in the barrel and haul it out and dump it out in the back.

I felt it was necessary to me being a woman working in the steel mill. I had to work and I had kids, they were my responsibility. I thought I did the best I could, and I think I did well. Well comparing to what I did, it was really hard labor.

Oh, the guys, I mean, I had no problem with the guys. They did look out for me and told me what to do, to be careful. They were very very caring, wanted me to be careful, had to careful and everything. I really believe that their spirit really helped me in my job.

Well, I mean that question there, it's not like, it was the situation that I needed a job. It was very, very important for me to have a job since my husband died. I had to work, I had no other choice. Well, the bad day was, you know, sometimes you have people would be doing sort of my job on maybe like on midnight. They pretty much don't clean it up. When I came in I would clean it up, shovel the chips and everything. So that's the only thing I don't like about the department. I felt that they should do their job as well as I do.

I thought that I considered myself a steelworker. Very much so since it's in the family. Around my father, brother and our uncles, basically all these people. You couldn't find better hardworking people than they are. So I felt that I belonged there and I looked forward to this particular job. It was hard work, but I was able to handle it. Well, they were all basically real good people. You know like Chris, most of them are gone, most moved somewhere and moved on, Deeno. I had, you know, the steel workers did their job. I thought they did it very well and I tried to be in that category. Well, I had to work shifts, well, it all depends. If I got laid off they would send me down to the Tin Mill, so I had to work three shifts then.

I think a lot. Maybe about four or five times I got laid off, I'm not sure, about that but I know I was laid off for a number of times. Oh, they send me down to the Boiler House, oh that was another job I did, geez.

In the Boiler House, cleaned, cleaned mens' room. Basically a lot of cleaning and I also did the blowpipe and soot. I wasn't supposed to be allowed to do that. So in the middle of the night I would go up there and would blow out the soot, because that wasn't supposed to be done during the day time.

Tin Mill, same thing, cleaned, shoveled, roll up the tin, put it in a bin. I did some work in the Tin Mill when I was younger. I worked in the Weirton Steel Mill. It was after the war, it was during the Korean War. I worked there, I was there almost six years or more. Well, I'm not sure, but I think it was the war involved, they hire women at that point in time. It could be the Korean War.

You know, cleaning and rolling up the tin, banded the tin.

You mean in Weirton Steel Mill, I was a tin flopper. Oh, flopped the tin and made sure they were real good and made sure there was any kind of flaws. It had to be perfect and then we just build it up so long and then they take it away. I mean this always it has to be a hundred percent perfect.

Well in those days, they didn't allow married women to work in the Tin Mill. Well, that's how it was then. Well, most women when they did work down there, tried to not let them know that they were married. I still have this letter that said I had to quit leave because I was going to get married. So basically if you got married, they didn't want you taking a job that a man could have. Yeah, it was like that.

You mean over the years that I was working in the mill? Well, I was very, very upset, with the layoff realizing that I was one of the people who had lost their jobs. I had two children going to school and it had been a very very difficult time financially. Having a home and trying to keep it up. Then finally, in the end, basically it went straight down and it's not been easy at all.

Women in the Mill

My last day,
well, the boss, I
also worked in
the Pipe Shop
and Bull Gang. I
cleaned every-
thing, cleaned
the   boiler,
washed down the
floor or shined
the floor and ev-
erything. Okay,
now, when they
told me I was
laid off. They
felt bad and did
not want to tell
me it was fin-
ished for me. It
wasn't like . . .
I'm sure that oth-
ers started to lose
their jobs at that
time. I had a
very difficult
time.

Realizing that as far as being a widow all these years and to have two
beautiful kids, good kids and it was very difficult for me not to be able to
help them like I wanted to do. So I had a hard time, it wasn't easy, it
wasn't easy at all. Nobody ever expected that to happen, considering
that J&L had great, you know, was known all over the world. They
decide they are either go elsewhere or just close it. I can never figure out
why.

Well, so many people had lost their jobs. They lost their homes and
the children and it has been very, very difficult for the majority of the
whole community. It wasn't a good life afterwards. Everybody has just
had a very, very difficult time trying to make a living and trying to have
a home and trying to help their kids, you know. I wish I was still
working down the mill. I wish I was still working.

Chapter 20

ဿ၇ၯ

Mechanical

H is appearance was that of a cagey, scholastic basketball guard, wiry and spry covering as the floor as he walked almost boyish in appearance for his years.

My name is Ronald D. Zuccaro and I am from Aliquippa, PA. I'm fifty years old and was born in Sewickley Valley Hospital. I have thirty-one years in the mill, my father is a retired J&L person also and he had forty-four years in the mill. I was hired on January 17, 1965 and am making roughly, fifteen dollars an hour, currently.

The Mechanical Department and Millwrights basically maintain the machinery, repair breakdowns, preventive maintenance, we would build up gear cases, pumps, valves, setting up roller bearings to put them on the trucks where they roll the steel on the mill. Basically, make sure that the units are running operable and any breakdowns, we repaired. In the maintenance department at one time we had probably 180, 190 people. Now, at the present time we are actually down to eighty people.

First job was a laborer. When I first started off I was caddying, before I got hired in the mill, as a caddie down at the Aliquippa Country Club, which was owned by J&L at the time. I use to caddie for a superintendent from the Tin Mill, and it was Paul DaVaney. Paul kind

Ronald Zuccaro

of took a liking to me and he was a very nice fellow and he knew that I was graduating from high school and that I was looking for a job. So he more or less got me my first job interview and put a good recommendation in for me, and I was hired in the mill in 1965.

From there I worked approximately seven months, then I was drafted into the military. I went two years into the military and came back with my same position as a laborer. Then there was a maintenance job open — extended D list. So I applied and was accepted as a Millwright helper, which I had to work as a helper for approximately six years.

I would assist the millwright, which would be to wash parts, run for tools, run for bolts and nuts and hold the flashlight for him, you know, clean up all the work for him — paying your dues. Then after six years you became a millwright. I became a millwright after a few tests that I had to go though to know how much knowledge I learned though the years. Passed all my tests and I became a millwright.

My best day in the mill would be when I had the opportunity to go visit my father working in the mill, the same turn I was, and walked down to see him. He worked in the Seamless Tube Department. He was

Blacksmith Shop

a tester operator, he had forty-four years in the mill. I just thought that it was very proud that I could go see my father working on his job and that he had a son with him, you know, at the time. He was very elated, as a father would be, happy to see his son. But it was an experience for me, it was a really happy experience. I finally got to realize what my father had to go through in his life to raise a family. I appreciated him quite more, knowing what he had to go through. After that point.

The worst day that I remember would be the day that I was laid off, the day that they told me that they were cutting back and I was out of a job. That was back in 1981 or '82. Prior to that I worked in the Tin Mill for sixteen years and then there was a bid that came up for the Round Mill, so I bid to go to the Round Mill as a millwright. They said in the Seamless Round Mill you would work forever because that's when the industry was drilling and using pipes. I took a shot and figured well, I'm going down there. So I did, and everything was goin'' along fine — all kind of money, making all kind of overtime, and all of the sudden the bottom fell out and I got laid off. Well, then I had enough time to carry me back to the Tin Mill as a temporary employee. Then I went to the

Byproduct, worked there, and then all of a sudden everything closed and I was out of work.

I was out of work for approximately two and a half years. In the meantime I took up jobs working at the golf course, drilling holes for the pin placements, doing odd jobs here and there, doing a little cement work here and there, trying to keep ends met. Finally they called me to go

Blast Furnace repair

to Warren, Ohio said there was a opening in Warren, Ohio for maintenance. So I commuted back and forth to Warren, Ohio for a year and a half, sixty-four miles each way, and everyday I did that for a year and a half.

Finally, they were looking for people down there at the Aliquippa Works. I was determined to come back here because I never took my severance pay which a lot of people did, and I said I'm not going to be out for a measly 4,000 dollars that they wanted to give me. So I just hung in there and then finally I had enough time to come back here at the Aliquippa Works and I started back here in 1989 or 1990 after the two and a half year layoff.

I remember, it would be the people, the people, the different ethnic groups, the immigrants, the hard workers and more family-oriented people

Millwrights

and closer ties. Actually it was a "shot and a beer" town which you would find a person that really worked hard all day and would go to the bar. You say, give me a shot and a beer, and he runs home and eats his dinner, more or less, goes to work in his garden when he got home. Basically, that is what I remember.

The people that I have worked with, who I have seen lose their homes from the layoffs and took their severance pay and couldn't find work afterwards. The struggles that they went through and the breakup of marriages because of that. That was the worst part, I hated to see happen. Oh, definitely, there was a lot of depression. Couldn't put food on the table, you know, and then the bills start coming in, which I experienced a little myself. Where you start to get a panicky feeling and you go into depression and then you start bickering with the wife, all the bills are coming in and you can't pay them, and you start getting a feeling uselessness, like why me, or whatever.

It's sad, like I said I've seen husbands and wives, who both had a job in the mill and they went and bought the extravagant homes, 70,000 or 80,000 dollars at the time, that's a lot of money. Then all of a sudden they both are out of work and their house is up for sheriff sale and everything they thought that was their American dream was kaput in just a matter of a year or two. That's the sad part.

Well, basically everybody that I was affiliated with at the golf course, when I use to caddie, was always a steelworker. I knew the higher class steel workers, which was the supervision part, which could afford caddies. I also knew the people that were just regular members that couldn't afford caddies. I remember them type of people. The shift work was no problem, but I think the hardest part of my family was the period before I was laid off at the time and I was out of work.

My last day, with the layoff, my heart was throbbing, wondering where I was going to go next. It really didn't dawn on you, you didn't really think, I'm going to be out of work now. Maybe that was for the best or something. Then when reality sets in and you realize that you have no paycheck coming in no more and you are forty-some years old and you got to go find another job. That's when reality sets in and you start to get panicky a little bit.

I would tell young people that the mill was a place that I provided for my family, but as far as security, I don't think you want to do it because everything comes to a cease sometimes, as I've seen steel mills do. I

Braising structural support

mean, as far as a secure job in the future, it's unpredictable, not like in the past when it was predictable.

You knew if your father worked there for forty-four years, you are basically going to work there for forty years and retire, but then you stop and think how long can this really last. That's the sudden part when reality sets in. I tell my kids that there is nothing secure in life that you can really go to unless you have that education, and you have a skill that is in demand at the time. Like if I would go to look for another job as a millwright, at my age it would be very difficult.

Roller scale repair

Chapter 21

ℰ⌘ℭ

Riggers

H e was amiable, engaging and jovial as he jostled you and talked much too loud for the surroundings, a gravely voice booming throughout the room unable to sense his own volume with hearing lost from metal industrial clatter. He was a compact fellow, a climber to heights of fifty to one hundred feet, still strong from years of hard work with a youthful appearance.

My name is Mitch Vignovich and I am seventy-two years old. I was born on November 19, 1923 in Aliquippa, Logstown. I was even born at home.

I worked for thirty-two years in the mill and retired fourteen years ago. I retired before I was sixty-two, they gave me that option and I grabbed it. That option was they paid you social security until you were sixty-two. Well they wanted the older fellows to retire. They were cutting the work force. Yeah, I was one of the first ones to go, they started in the early 70's.

I worked with the Rigger Department. They did maintenance, mostly maintenance, or we put up buildings or repaired the blast furnace when it went down. Out of all the maintenance they started the job when there was a big job and then the other ones fell in like carpenters. But the

Mitch Vignovich

riggers, I believe they were number one when it comes to maintenance, and then the rest of them just helped out.

I think the biggest job was when they took the blast furnace down, all the way down. They took it all the way down, all the way from the top — the big bell, the small bell, the bleeders, everything. I think it took at least a year and a half. They took it down for maintenance.

It happened when the salamander inside the furnace got to big. The salamander was the steel that was on the bottom of the furnace. When it got too much it would cool, it was too thick, it was maybe ten, twelve foot thick, so that was when they had to clean the furnace out.

Yeah, there was a lot of emergency work. They called us out a lot of times at night for emergency jobs. We burned and rigged, put up the Annex Building down in Seamless, that was one of the big jobs. So basically we were an internal construction outfit, and moving heavy equipment.

I enjoyed everyday, I enjoyed the riggers until the last couple of years, you know. I enjoyed working in the riggers. Yeah, it was hard work, but it was good work. The fellows and the job, it was always a challenge. I enjoyed being on big jobs. Well, the big bell on the Blast Furnace, that's South Mill when they started the Blast Furnace. North Mill riggers always had the top — they took the bell down and put the bell up. And the South Mill riggers, when they just start working on the Blast Furnace they did the dirty jobs. They changed rails, did most of the bad jobs, and down in the stockhouse they gave us the dirty jobs. I was with the South Mill Riggers until the last ten years. We came up north, everybody came up north then we worked on the top in the last ten years.

I think one of the worst days was changing a motor in a coil kneeling fan up on the roof. The fan's up on the roof in July. It was hot and dusty, it's one of the worst. You had to . . . you had to use a ratchet, you had to lower the motor down with a hoist — that was one of the dirtiest and worst jobs I hated. Well, they have to be changed, and the only time they change them is in July when the cranemen complain, you know. They didn't complain in the wintertime, they complained in the summertime, when it was hot, and that was when you had to change it. Not only hot, it was the dust.

I'll tell ya, they called us out when there were a lot of break downs, say a crane came off the rail or something. Where you had to use a crane to pick it up and jack it up, yeah, they were all, bad, you know. I enjoyed it, but like I mentioned there was the coil kneeling fan in July. The coil kneeling department, you know, the fan's up on top. That's in the Tin Mill, where they kneeled tin.

I remember Joe Rosa, he was a smart one. Brownie Smojanic, he was a good worker, Fred Vignovich, Lubar Vukmaravich, Matthew Vukmirovic. They were good guys, good workers. That was important in the mill, being a good worker. You didn't mind going out, no matter how bad the job was, if you had a good gang, say, three or four in a gang that were good workers, you didn't mind. I don't care how bad the job was, as long as you had good buddies working with you the day went fast. Like changing girders down the Seamless, you know, we changed all the girders, that was a bad job, too. The coil kneeling was bad, that motor, 'cause there were only two guys that were able to get in that little space on top where the fan situated up on the roof. Two men. That dust, when you hit something that dust would all come down. Oh yeah, Sinter Plant was another bad place.

That was the dirtiest place in the mill, I think. The Sinter Plant. You know, off hand, there was always some kind of big job there and that was the dirtiest place, and you know, for when it concerned you know, say, the plant, it was one of the dirtiest places in the mill.

You know what, I never thought the mill would shut down, the Aliquippa Works. Higher supervision and a lot of politics, but I think J&L had one of the best maintenance outfits. When it came down to it, they were good. I can't remember my last day in the mill. I was glad to get out the last couple of years I was glad to get out. I had enough. I think it changed. I don't know if it . . . I think everybody, the workers or something changed because, for some reason, it changed. I don't know, for me it did, but before that for thirty years I enjoyed every bit of it. Maybe it was myself, I don't know, a lot of the new fellows came in, they weren't up to it.

Some of the younger guys wouldn't listen. There were a few good ones, there were some, but some of them you tell them something . . . Like I told one go get a prying bar or something — I forgot what it was — he came back, we needed it at the moment, but by the time he came back we had already had that job done. When he came back he said, don't ever send me for nothing else like that. I went after a hundred of them and when I came back the job was done, not a hundred, 500 times. You know, smart asses like that.

Yeah, Fred Vignovich, I think he was one of the best workers and best climbers in the mill. Yeah, they did a lot of climbing. One day they cleaned the struts and they cleaned the shields. Well, South Mill never did that, but North Mill did that, they cleaned the shields and the trusses in the Bessemer, and cleaned the runway from that black scale. You know, that was one of their weekly jobs, when the Bessemer was in operation. Yeah, that was bad work.

Well, like in the Byproducts Department they used to send riggers in the furnace, in Byproducts to put scaffolding up in there for the bricklayers, or to shore the brick up. His feet were burning and the guys outside would holler, Fred, your feet are burning your shoes are burning. He would still stay in there. Yeah, it was dangerous, it was all dangerous. I'll tell ya, you know, and it was not only even a smart rigger, I think he was smart rigger, too. Him, Curley Verez, there were a bunch, Red Patterson, there were some good ones.

It was dangerous. Well, this one instance I can remember when a crane came off in the slab yard between the small slab yard in between the Strip Mill and the Blooming Mill. They called us out, we had to put the crane back on, and it bent a rail so we were working midnight, and

Blast Furnace

the rail was bent over hanging. One of the riggers, Brownie Smoljanic was a good worker, too, he went up there. The foreman told him to go up there and burn that rail. The rail was bent like, it was sticking out over top. He went up there and I seen him when he was ready to burn it, and I hollered up there, hey Brownie, leave that go, don't burn that.

This is true, too, and I told him don't burn it, you know, leave that go, I'm coming up. So when I went up there he was sitting on the rail. If he would of burnt it that would of threw him fifty feet in the air. So when I came up I grabbed the torch out of his hand and said watch, and I stuck my hand out from the roof and I burnt the rail, and when I burnt, that rail went down like a spring. If he would've been sitting up there — you know, he was a good worker but you got watch your other. That was one instance I will never forget. He remembers it, too.

I wasn't there when Matt died, but from what I hear they were burning down a crane cab in the slab yard and he was burning part of it. But on the same token, just like previously like I told you before, like I was watching. There are older riggers that should've been watching him and having the right hitch on the cab, but when he burned the last piece down it fell over and he came down off it. But, I think, who the heck . . . you can't blame — I don't know who was on it cause I wasn't on it. But when a burner is doing his job he can't see, he got the goggles, he's concentrating on what he's burning. There are a lot of times where you got these other riggers that are standing around, should be watching him.

Well, on that case, they should have had a good hitch on the crane, they should've been watching him, I don't know. I'll tell you what, it's the riggers that were sitting up on the crane girder, they should've been watching it. He's down in that cab, I think they had a lot to be blamed about, just like I told you about that . . .

Definitely, believe me, as long as he knows what he's doing. He was a good foreman, he never got over it but, hey, a lot of time the foremen, the workers, the foreman can't be there always. We worked well with our supervision in the Riggers, we worked side by side. There were some good riggers and good bosses. Yeah, the foreman, yeah, like Johnny Parducci was one of the best foreman, Steve Leechman was a good general foreman. South Mill, they was always saying Steve Lechman being the best, or his uncle Lechman up North Mill.

View from top of furnace

Small bell

Base of No. 4 Blast Furnace

Fred Vignovich – End of strike in 1959

Steel erection

Girder erection with portable crane

Chapter 22

ℰ)ℭ

Electric Crane

He seemed quieter, more demure than his nickname — Spider — implied as a rough and ready member of the Bull Gang, a diverse group of men who erected and repaired the overhead electric crane — the workhorse of the steel mill.

My name is Dan Kosanovich, and I'm from Aliquippa. I'm seventy-four years old and was born on the South Side.

I worked thirty-nine years and six months I believe, in 1941. I think we were making only three dollars and seventy cents a day in 1941. I retired in 1980, August 1, 1980. I was making about, pretty close to twelve dollars an hour then.

I worked in the electrical crane repair, it was called the Bull Gang, in the mill. We repaired all of the cranes throughout the mill. We didn't work in one particular department, we covered the whole mill. All the cranes, and if they needed repairs, we repaired them. The Bull Gang, its function. The Bull Gang guy is a part of a millwright, he is part of a rigger, and he is part of a boilermaker, and that's what his job consists of. A little bit of everything, like when something was broke down and we had go change a wheel on a crane, or working in the Soaking Pits.

Dan "Spider" Kosanovich

We got to work about seven in the morning and we got our assignments. We usually got our assignments by telephone calls, about break downs, and some days we got sent to a department. About five or six men would wait for a break down. When it comes about everybody goes on that job, repairs, and gets going, usually in the Production Department. They want them cranes right away, and they sort of push you on that.

All overhead cranes, yeah. The overhead ladle cranes, they picked up. They claimed to be 150 tons, but they picked up more than 150 tons. It ran on tracks and it had two trolleys on it and it had four girders. Two large girders for the big trolley and two small girders for a smaller trolley, used to hook to tilt the ladles that had the steel in them.

I started out as a laborer, first in J&L, and I moved onto the Cinder Plant and we used to cinder ore. From there I moved to the Tin Mill, and I worked as a packer in the Tin Mill. At that time I made up to five dollars and eighty cents a day at that time, that was 1946 or 1947. From there, I was a coil handler in the Tin Mill.

Then, I was in the service. I was gone for three years. I was gone from 1943 to 1946. The mill ran pretty good, because a lot of the women went to work there and they did a wonderful job for the steel mills. A lot of women became crane men, and they run them cranes. The older men stayed back and repaired them. Then I was in the Tin Mill and moved into the Electrical Crane Department, because I had a MOS, like a I had knowledge of maintenance. That helped me get into the Electrical Crane Repair. From there — I worked there quite awhile, thirty some years — and then the last five years I ended up being a crane inspector. I would go about inspecting the cranes, checking the shafting and the hoisting and the cables, and making sure everything was safe. If anything wasn't safe, I would make a report out everyday on the crane. I wasn't allowed to inspect no more than three cranes a day, because you can't get over. You can't do a job well, doing more than three cranes a day. Yeah, it was safe down there as far as the cranes were concerned, but if anything was unsafe, then I marked it on my report. It went in the red pencil.

One report went to the department, one report went to my foreman, and one report stayed with me. Because if I would get called on the carpet or anything, I would bring my report out, get my boss to get his report out, and compare it with the department. Well see, I was inspecting a barge loading crane, and one of the girders was losing its rivets, and one of the bridge motors was cracking away from the girder. I kept writing it up and writing it up and my boss called me in and he said, you know, there's a motor cracked on the girder down the barge landing and the girder's coming loose. I said, yeah, but look at my report. He pulled it out and it was marked in red and he said, good, no problem.

I was three years old when we moved into Aliquippa, and it wasn't Aliquippa then. It was Woodlawn and I grew up on Plan 11. 1924, and I can't tell you too much about it because I was only three years old. What do I remember when I was three? But as I grow older it was a real nice place to live, up there. We had all kind of neighbors, we had Serbians, Italians, Russians, and the colored people, and we all got along real well. In those days, you could leave your door open and nobody would bother you when you went to bed at night. It was wonderful living in those days. Today you can't do that. So Woodlawn was a planned community, it was a company town.

Each Plan was set up by the company. There was Plan 1, Plan 2 all the way to Plan 12. Plan 6 is right above us right now. Most of your foreman lived up on Plan 6. Plan 1 was back towards West Aliquippa. Plan 2 was Logstown. Plan 3 was going up before you get to the police station down there, municipal building, went up the hill. Then Plan 4, I can't remember, Plan 5, I can't remember, Plan 6 was up on the hill, Plan 7 was going up the hill towards McDonald Heights, Plan 8, 9, and 10 I don't know. There was Plan 11 and Plan 11 extension. There was two of them, and then Plan 12 was up where the high school is.

The best days I had in the mill was when I got to be a crane inspector, because I worked alone. I didn't have no foreman and all I had to do was go and inspect my cranes and make reports out every day. Then we would have safety meetings maybe once a month. Sit down and we'd talk about a half an hour once a month about safety because the job was really treacherous.

You were always in oil and grease or in heat and cold. See, our job, maybe you were in a hot place for four or five hours and then another job broke down and you went outside and maybe it would've been zero out there. Then you went from one hot job to a cold job, or a cold job to a hot job — a lot of temperature extremes.

Well, the worst day was when it got to be ninety degrees outside and then you were working in the Soaking Pits and them doors opened up, below you. It's almost 150 degrees without the pits being opened up, and then when they opened up they're hotter yet in there. So the Soaking Pits were in the Blooming Mill where they took the ingots and they were heating them up again.

I wore winter underwear all year long because when you sweat in the summer time and the sweat, you hang onto it, it cools you off. All of us millwrights wore winter underwear all year long. I remember working with the guys, it was a good bunch of guys that we worked with. We had no problems — arguments just like anybody else would have arguments, but they didn't amount to nothing.

No, I really enjoyed working for the company. J&L was a wonderful family company to work for. They were very, very good, they were good people. I took an early retirement in 1980. No, LTV was there when I took it. Things started to change when LTV came in.

Well, the plant started going down little by little. I talked to one of the directors down at the main office and asked him about retirement. He said that I was fifty-eight and a half years old and things don't look good. He said, if I was you, he said, it's not going to hurt your pension or social security at all. They are offering you 400.00 dollars a month. He said, if I was you, I would take it. So I took it, and was happy that I took it because after that, everything went down after that.

Really, they moved quite a few, the mill to Cleveland and Indiana Harbor in Indiana. The only thing left is the Tin Mill down there, and they're still producing and making money for the company. They sold the 14 Inch Mill to a couple of their executives and they're making all kind of money with their 14 Inch Mill. These guys are going to expand.

Oh yeah, I remember quite a few of them. I worked quite a bit with Nick Yurich and Emil Paige. They sort of kept us Serbian guys together. I worked the night that Matt fell from the crane when he was burning rivets out of the walkway. The solution that we come up with, the painters were supposed to clean that runway off so they could see how many rivets were missing out of it. It wasn't done, and this is the big reason that the walkway give away when Matt was burning there. Because he didn't know how many rivets were missing out of there. When he burned that last rivet out and the rivets behind that he thought were in, weren't in, and the walkway and him all went down together . . .

Yeah, the painters were supposed to clean it up that day and they didn't clean it up. The four o'clock shift come out and they didn't clean it up. Yeah, you had people on either side of him. They were watching, but they didn't know either, you know. Yeah, that was a sad night. Your dad was a nice man, a good worker like most Serbians.

Not at all, my family wasn't affected. They accepted what I was doing down there. Well, not too much. We in the Bull Gang, most of our gang was out on daylight all the time, and then we would have four or five men, say, in the North Mills, about four men in the Seamless Tube, and four in the South Mill. You had about twelve men during the four to midnight shift and the midnight shift. If anything big broke down then they would pull all twelve men together. But the small stuff like in the North Mills, the four men would handle it. If anything large broke down, the telephone calls would go out and all twelve men would pull together.

My last day was wonderful! My last day, the foreman told me don't even change clothes, you go around and visit everybody. Shake hands with them and we'll have coffee at noontime and a piece of cake for you and then we'll let you go.

What I would tell the young people, I would tell them hey, if you have a chance, get an education and go to college and stay the hell out of the steel mill, because it's hard work. You know a lot of people say this guy was sleeping down there, but what he was doing when he wasn't sleeping, it was dangerous. I would say to the young people hey, if you have a chance, the steel mill pays good money, too. Today they do. But hey, if you have a chance to get a job someplace else with an education, go out and get it.

No, I'll tell you that I really enjoyed working for J&L. It was a wonderful company to work for. But when LTV came in, things just changed drastically. I don't think they had the knowledge of steel that J&L had.

Plan 6

Portable crane

Ore yard overhead crane

Erecting crane

Overhead crane

Section V

Service

Chapter 23

§∂)C∂

Clerical

S he was proud to have been part of the steel effort - a demure featured feminine face in a sea of rougher hewed, masculine personas, speaking of IBM punch cards not hot iron, yet possessing the same conviction and spirit.

My name is Ann Baljak, and I am over sixty-five years old, not much, you know. I worked for thirty-one years and was hired in October 1951. I was hired as a typist/clerk in the Tin Mill office and made one dollar and forty-six cents an hour. I worked until 1981 — no '82 — then they let us go and gave us an extra four hundred dollars as early retirement.

We worked for the Production Planning Department where we dealt with orders and processed orders through the Production Planning Department, most of the time. This was done by hand and was not computerized. We typed with carbon paper for duplicates, and used an old Gestetnner to reproduce copies. No, I don't know what it was, it used that ink jelly stuff. I can't remember the name of it. It was called a Hectograph, I think. Everything then converted to Xerox, and we copied everything onto cards and then to IBM.

Then we put all our information on our IBM cards and transferred them to the keypunch cards and we got the reports back and had to check

Ann Baljak

everything. We had to verify sizes of orders, companies to be shipped to, what they ordered, how much, were they accurate. We had to be accurate.

The best days in the mill were when I started. I was twenty-one years old when I first started working there. I had to apply at the employment department and then my one friend, so-and-so knows so-and-so and I think that's how I got in.

I remember the day Kennedy died, that was a very sad 'cause people who were working, we were listening to the radio, but that wasn't allowed at that time. Some had to wait till they got home. We basically weren't allowed to listen to the radio. Some people found out from the mill when they called their families.

The work environment was okay, you know, the lighting wasn't the best, and we had old typewriters. The supervision was great when I started and we worked hard, and then over the years you could see the decline in the supervision and the lack of rapport with the workers. I don't think there was a decline in the supervisory personnel over the years. I think anybody can attest to this, the United States didn't want to make steel anymore.

So they were blaming everything on the unions, and it was the union and the employees, and they just let it happen. The supervisors, you could just see the lax in their authority, they didn't care anymore. The unions input to this was nothing, they really didn't do much because it was coming. I mean, they tried to fight, they tried. Before we were union members, if you got caught sleeping on the job you got fired, but after that you could stay. It just became lax. You just seen it coming.

This didn't come from ground level supervisors, it was from the top. When we could see that the supervision of departments were, well, they just didn't care. We knew something had happened when they had to hire all the minorities, and they got away with murder. So then, why should we work the same wage they're working for, they weren't doin' nothing. We didn't think this way until after we left the mill. We didn't know what was going on. The minority hiring, I mean, was a situation where they didn't get any work for forty years and they had to sort of make that better for some people. They had to hire anybody off the street. They weren't qualified working there, but they had to have a certain percentage.

The good times are memorable from the early days, 1951-52. We used to have noon hour and Christmas parties at lunchtime over the years. We had everybody pitch in and they brought food in, we also had a bowling league. My early days in the mill were fun, they really were. You looked forward to when you had to go to work in the mill. And then when this started, you know, I would say the last ten years that I worked. I think the change started in the early 1970's. My last days in the mill are what I would like to forget the most. This is when the early retirement program first came about. You have thirty years, you're not old enough. You had to have the magic number of seventy and I had that, okay I had that. We all had thirteen weeks over the year so I had that. That was another ruination of the mill — the thirteen week vacation.

Yeah, because it was suppose to create jobs for other people, but all that happened was that we worked double shifts to cover, so it never really created any new jobs. That was the downfall, 'cause people had thirteen weeks vacations, so when they went back to work they didn't want to work. So it all jelled, they knew this was going to happen — lax management and lax employees. Then they wouldn't tell us if they were going to lay us off. So there you are hanging, you don't know what's going to happen. I worked in the office all my life, so if they're cutting back and they could send me down to the mill to work, which I never went.

We walked through the mill to get to our office. We just had to be careful 'cause there were overhead cranes. You had to be alert. We had to wear helmets going in and out of the mill, after a while.

I remember my boss, Ed Davis from Monaca. His sons are professional basketball players. He helped me through a lot of hard times. Definitely, there was a lot of camaraderie there. Like I said, the early days were fun, were great, we had retirement parties like we talked about. It was just great and then, after a while it wasn't so great. Well, I liked the fact that our office was big, it was huge, there were like close to a hundred people in it. We saw the same people everyday. You know, most of them were men, but we had about twenty women that worked. But we fit right in, there were no slurs or no dirty talking. They respected us. We were afraid to tell the management or try to change anything.

There were no open discussions of problems. Well, the ones that did open their mouths, they were troublemakers. They would shift around, well, we didn't have that in our office. I'm telling you we had a nice atmosphere. Those were the happy times, we were dedicated workers, we worked eight to five. Religiously, we were there everyday and there was no goofing off or reporting off. We didn't have that there. Dedicated workers were common in the plant in the beginning. In those days toward the end, it wasn't, they reported off if somebody sneezed.

I would tell anybody just do your job and be conscientious about what you're doing, be proud of what you're doing. When I lived in West Aliquippa, which is a little cocoon — I don't know if you know about West Aliquippa. One way in, one way out in Ripleys believe it or not. Home of Henry Mancini. Right. Right. So I really had a good career. It was just those last years where they wouldn't tell the mill was going down as they were shutting down. Tin Mill is still working, by the way.

Clerical staff

Office staff

Chapter 24

ℰℭ

Union

H is optimism and cheerful smile was pervasive as he related the past glory of the USW-1211, one of the largest union organizations in its day. Yet, he seemed hopefully optimistic for future workers who could remember the lessons of the past.

My name is Melvin Kosanovich and I work for the union at 501 Franklin Avenue, Aliquippa, PA 15001. I am sixty-four years old and was born in Aliquippa. I was hired in 1951 and have been a general griever around twenty, probably twenty-six years in the mill. I got forty-six years in the mill. I'm not retired, I still work, still active. I'm the oldest active employee on the board.

My first job in the mill was laborer at the Seamless Tube. I stayed there one year. Believe it or not, at that time we had around 14,000 employees and they were pretty hot on sports, so there was a basketball league in the mill. The Carpenter Shop was where I was looking to go, because it was steady daylight and they had a good basketball team. So the guy from the Carpenter Shop who I knew ran the team came down to the Seamless and recruited me and another guy to go to the Carpenter Shop. That's how it came about one year later.

You see, at the time, they grouped you in departments. All the Serbians went to the Riggers, the Anglos to the Rod and Wire, Greeks to

Melvin Kosanovich

the Paint Shop, Italians to the Carpenter Shop, Afro-Americans to the By-products, Blast Furnace and 14 Inch Mill. I worked in the Carpenter Shop for maybe twenty-five, thirty years. The Carpenter Shop built platforms and scaffolds for people who had to fix things. We were . . . we did the whole platform. We were no special department, we worked from north to south mill. So that other people could go on these platforms to repair those jobs.

Some of it was putting up pipe scaffold, which is a five foot section plank on top of it. Some of it is hanging with cable or some was put inside crane girders. It was just where we could get the platform for people to repair the equipment.

My job title was Carpenter (Class A). The Carpenter Shop had A, B, and we had apprentice classes. I went through an apprenticeship for two years, and upon completion of that and going to school you became a C Carpenter. After a period of a year you became a B. Final year, after five years you became an A. When I was hired I made one dollar and fifty-one cents an hour. I am considered right now an hourly employee. I am making around fourteen dollars an hour, fourteen or fifteen.

Young Melvin Kosanovich

A general griever job is more or less like the lawyer for the union. People file grievances and complaints, it goes in a first step, second step, which is not my area. Third step, is a full fledged hearing of myself . . . let me explain.

The first step is the man himself, he has to write a grievance, it's a complaint. Second step goes to the committeeman of that department, once he hears it and it's denied, it comes to me for third step. The committeeman is an elected official and he is elected to that job down there. If he thought it had merit and he presented it to the company, and they denied it, then it comes to me.

The third step process is the full fledged hearing with myself, the committeeman, and the witness, the company lawyer and the supervisor who the incident occurred with. One time, we had a stenographer, later on we had a tape recorder and now, because we're so reduced, we just tape it by writing minutes out ourself. We were doing nine a week, in fact, we had two general grievers and two research people. It's really been narrowed down to maybe three or six cases a month.

Right now, we got about 650 employees working in the Tin Mill. Well that, and we have general maintenance is still running. But that's all part of one, that's inclusive. When I was hired or at the maximum we had 14,000 employees. This was a typical regular grievance process. Say a guy is a one year senior to another employee, and we have a list that said you provide overtime to the oldest man. Now they go to you, junior to me, for overtime, and I find out.

I go to file myself in the First Step of the grievance process. I say I was denied overtime by my seniority, which is Section 10 of the contract. They'll deny it, or sometimes answer it and say yes. If I'm denied I go to second step. Second Step grievance is with the committeeman and the superintendent of that department or a representative. If it is denied there, which it normally does, it will go to Third Step, which is general grievers part of it. If it is denied there, Fourth Step is the district staff rep, and if it's denied there it can go to the Fourth Step or maximum arbitration, which is the lawyer from the company, the union, the staffman and myself, and come to an agreement. Those are the five steps that we have.

They rarely go to that fifth step cause it's very costly, it probably costs in the area of around 3,000 bucks total if you go to that step. We normally try to resolve it in a third and fourth step proceeding, because if you take too many to that stage you're broke.

The best day I can recall was when we had a guy who had one eye, and the union term was monocular vision. The company called him "cocky," one eye guy. This was maybe fifteen years ago, and prior to our administration they wouldn't take the case to max arbitration. The guy came up here and we saw him working on . . . he was a pipe fitter who was working on poles at Ambridge football stadium up in the air maybe fifty or sixty, maybe seventy feet. So what we did, we hired a doctor of optometry, we hired a regular doctor, and they are claiming that the guy had no stereoscopic vision so he couldn't see depth and might wind up killing himself. So we took it there, and after around a week of testimony we proved that the guy did work outside, his monocular vision didn't affect his climbing, and they gave him his job back.

That's one, let me give you another big one. We were . . . when me and the past president were in charge, they were going to close down our repair shop. The repair shop is the people who fix motors. The company went out to General Electric and bought maybe 500 new motors and told them they could fix them all. In the meantime we had a place here to fix them. They shut our place down and we took them to arbitration, we took them through the 1st, 2nd, 3rd, and the 4th steps, as the procedure goes. We took them to max arbitration and, believe it or not, after they closed the place down and spent 500,000 dollars on the motors, we made them open it back up and hire the people back and abide by the contract that it is. That was the biggest victory that I ever had in my life, very proud of that decision and that was around maybe twenty-five years ago.

I think that they wouldn't want them to lay these guys off and hire someone out of the streets who knew nothing about our mill, who couldn't fix our motors. Our people were twenty-five, thirty, forty - year veterans who could do that specific work. It was pretty much, their deal at that time, there was a lot of corruption going on, so they gave a big job out to GE. But our workers could do it better than anyone out there cause they knew the job. I think that's what they would decide.

Probably the worst day was when we had a couple of people who were stealing, and one in particular was stealing from a Children's Hospital fund and one was stealing from a . . . I think it was a coffee club. The two departments came to my office and said, if you let this guy back you're a loser, we'll get rid of you. So thinking about that, they were trying to tell me don't do my job so well. We had arbitration and, lo and behold, they let both guys come back. I was . . . I did my job but I didn't care if he came back or not because it was two stealing incidents. After that the members picketed the halls and made phone calls and threatening phone calls for doing my job. It was one of the worst times of my life because we won the case, which I really did care if we won or not.

The union had very much power, a lot of power at times, too much power because most of the things we done we had a lot of power. We had a lot of hiring. Whatever we wanted they gave us cause they were making money. Things got slacked and they couldn't afford to pay those things, but yes, we had too much power. At times I think the company had too much power, but I think union had more power. They were a powerful organization. When the president of Local walked down the street people just moved aside. He was the main man of town, he was like the mayor of the town, big time guy. 14,000 people, it's like a whole town, big, strong guy always. Union president? Normally you would stay union president for two terms, which is six years.

I would like to forget that they tore the place down. I'll tell ya, right now, when I was hired here there were 14,000 people and now we are down to 650 and you see the plant now, it's sickening. Saddening to see. It's leveled completely and there's not a stitch of industrial work there other than Tin Mill. Knowing what I know now and being around for forty-six years in the plant, I know that a lot of the departments they could've done something about it to keep running, and they didn't. I think the company chose to put the money in their pocket rather than repair the plant. They let it deteriorate to the point where it just fell apart. Because they just wanted to be rid of it. They didn't want to have nothing to do with ESP. ESP is employee stock ownership, and they did not want to sell it to the employees. They didn't want to do that. They wanted to level it, sell the scrap, and put that in their pocket and go from there. They did not want the plant at all.

There was a guy named Blackie DeSena who was union president here. That guy, when he was here he was straight, hard charging, knowledgeable union president who knew as much as anybody in the business and would fight for four step cases. Fought for pennies. Because he thought he was right and as far as right, now goes, I think the president right now is a very compassionate guy, hardworking guy, and is probably as good as any worker here and a union man that we have had.

A union man is who first of all sets an example, and you look on the product and see that it is union made. If you go to buy a car you see if it is union made. In the US, you go pick up pants, shoes, shirts, you want to look at union label. In my opinion if you buy American stuff, that's a union man and that's me through and through to the day I die.

Well, initially, let me go back to my mother and father. My mother and father are both Serbian. My mother was born in Pittsburgh, my father, believe it or not, when he came here he was like a rep for the Serbian people because he could read and write Serbian which most of them couldn't do. So he was a union man and there were times he would sign people to join the union in 1936, and I recall one time when he went to work and he was signing up people, the company knew about it. They were going to go to his locker and before he got there a supervisor, who was Serbian, who was company man, got to him before they got there and said, get to your locker, take them cards out of there because they're going to come and fire you. He took them out of there and when they searched my father's locker, they weren't there. But two days later he was signing up employees. That's what I remember about my father, many picket lines. Many times my father came with bruises because he was a strong union man, very strong.

The union was formed in 1936. I was only five years old. I recall sitting on a picket line with my father and they would bring me sandwiches, make coffee, and there wasn't anyone who was going to win with an old-timer. They would go out for what they thought was right. They would go out at the drop of a hat back then, and they thought the union meant striking and that's what they did.

The 1936 strike that was when they were going to take some wages and benefits off of them, and they didn't stand for that. I think they stayed out around four months. I think they restored all the benefits and then they went back to work.

The next one I recall was in 1959. We were out for 116 days and that was over the same thing — wage reduction and benefit reduction. We had a strike fund here which wasn't too plentiful and we give some needy people some help with the food and utilities and their rent money, but not everybody. Most had their wives working or had some other kind of job, got a little job someplace and they had a couple bucks from that. We would try to take care of the needy, but even the people who have money always got in line to get the stuff that the poor people got. There are always greedy people around on both sides of the table. That was the last one. That was the last one that I recall — 116 days.

If I would change anything here, I would, like I went Tokyo, Japan and I saw how they worked. They work it where the company and the union are in one office together everyday of the year and the union people are paid by the company. There are no grievances because whatever issues come up, they are there to resolve it that particular day of that week. I think that's the system we should be, in my opinion. I like what they did over there, 'cause I asked a Japanese worker how many grievances he had and he said, what is a grievance, and that stunned me because they are resolved when it happens right there in that office and they are all full-time and work together. The superintendent and the president in one office.

It certainly would have changed things 'cause you would've resolved the issue before they became big problems. I don't think there'd be no shut down because they would've agreed to . . . if they're shut down like they do in Japan they could place employees in other plants. That's what they did over there. They always wanted to be company and they knew we were union. They would always say they want it to be together, but they would never do that, the wall was too thick between us two.

The last day when we shut down was very saddening, heartbreaking just to see. We knew that the people coming out of that gate would be the last time, and being around as long as I did it made it difficult to see. Tough to watch. Yes, well, they gave them notice, you had to give ninety day notice, and that particular time we were bringing in the people

from outside contractors to level the land to sell the scrap. That took about three years. We made an agreement that if they would recontract for demolition in here, which we wanted them to do, they would hire our craft people to help, who wanted to help. If they wanted to retire we let them retire. Sure, because it was either that or layoff or a pension, and most of them weren't

United Steelworkers of America – Local 1211 Hall

eligible for pensions at that time. I was a craftman so I know the craftman's carpet tool, I know it myself. We had an early retirement plan which we called seventy-eighty pension. If a guy had those numbers — age, and years of service — he could go on pension but he had to be fifty-five years old it was at that time. So if the worker was fifty-five years of age or older and had at least fifteen years of service, he could retire with the minimum pension. But now there is no more age limits allowed, age is gone, it's just thirty years. The purpose of that, is to get younger people in the mill, that's one of them. And to give people a chance to work, I think that's why.

I would tell the young folks that it is a good place to work, it's a good wage, good benefit, and a good place to provide for your family.

No, I appreciate the interview, I think it should be told by somebody, and I'm glad to be a part of it. I'm glad you're doing it and I thank you for it.

We had good workers in that mill, Matt Vukmirovic, Fred Vignovich, Nick Lackovich, and Regis Schultz. Those four, they could do anything. You wanted a man like Matt in your gang cause he would do most of the work. Stand there giving more hose or more oxygen, get him a cup of coffee and he would burn all day, sparks and dust flying everywhere. He was quiet though, only smiled every once and a while, always looked like he was thinking about something else. He was a hard worker.

Union Hall

Union flag salute

Listening intently

Union meeting – The vote

He sat in the old union hall, the building now sold, a back entrance used to access the dark and dusty office confines. A far site from the dozens of personnel in the old days - one secretary and a few officers. The formality was gone short-sleeve sport shirts were acceptable work attire, as he reminisced of the past formality lost.

My name is Richard Vallecorsa. I'm fifty-nine years old and was born in Aliquippa, Plan 11, in a house. Plan 11 when I was a kid was made up of Italians, Lebanese, Blacks, Serbs, Croatians and Czechs, and the regular-type people, but I say that not in a bad sense.

The plan system, I guess, was designed back in the early 1900's where they had a Plan 1, Plan 2, Plan 3, Plan 4, and they would build houses and square them up, survey them. Those plans were designed and most of them had company houses on them. People rented or bought them later from the company. They were like segregated, with that not actually being on purpose. Plan 6 was the bosses, West Aliquippa was workers, and Logstown was workers. Plan 11, Plan 12 was more high class then Plan 6. I don't think it was a bad idea for that period of time.

The Mill

Richard Vallecorsa

I was hired on August 18, 1954, eight days past eighteen years old. I was hired into the Welded Tube. Well, we just past August 18th. This is what, '95 I'm in my 42nd year. I think what I started at was like one dollar and sixty-eight cents an hour.

I worked from '54 through '59 in the Welded Tube, I worked at the Buttweld Furnace where you used to make pipe, one piece at a time. Then they put a Continuous Weld in where they came in fast. Where we used to have sixty to seventy men on a turn, we went to twenty-one turns. So from there I went to the Shipping Yard where they used to put the pipe in the gondola cars for the railroad and ship them out, and I put in for apprentice in 1959.

I went to General Maintenance and was a Pipefitter Apprentice and I stayed with the Pipefitter in the General Maintenance until 1962. Then I went to the Tin Mills as a pipefitter, once I got my rate to Tin Mill. That was the way it was, wherever you were needed when you got done, they shipped you there. So I went to the Tin Mill and I'm still like a Tin Mill employee.

However, the way I got into the union was interesting . . . Well, usually, let me see what happened. I guess I was always a little bit interested in the union. I used to talk to my dad on who should I vote for. I wasn't too hip on what was going on, but he used to tell me who the good guys were and to vote for this guy or that guy. I always made sure I voted, and I don't know, back in the late 60's early 70's I started going to meetings and got involved. When we used to have committees I was like on the hospital committee. We use to go visit the guys in the hospital. At that time we used to bring them cigarettes, you know, or a box of candy, you know.

From there I went and I think somebody quit or died or something, as a guide or a guard, back in 1968 or '70. Sometime back then. I was interested, so they appointed me to that job to fill the term. I was that for a while and then I ran for trustee, which was . . . I think I was there for like six years as a trustee. Then in '76 I think it was, I ran for treasurer and I held that position for nine years, three consecutive terms, and then I was elected vice president one term, and then I've been president since 1985

Yeah, ten years. I'm the longest serving president from United Steelworkers Local 1211. There were over 12,000 union members during the war, it might have been like 14,000 or 15,000. But even as late as '80, '81 was 12,000.

In Pennsylvania, we might've been number one but I think in the country we were as high as number four. There were a couple of plants that were larger, it didn't really mean that much. But we were big. Out of 8,000 Locals we were in the top ten.

What I do now, I'm the head negotiator between me and the company on problems that exist in Aliquippa. When we go to negotiation, which will be reopening here pretty soon, I represent the Aliquippa Works, and I'll vote on the contract before we bring it back to the guys. I'm also a General Griever like Melvin. Even though he does it, I can do it, too. So we resolve issues between the company and the union all the time. Retirees, I take care of some of their stuff when they come down here — insurance problems, somebody dies, sometimes they'll call us, you know, spouses.

Why it changed before, when the president was like strict. You know, you would knock on the door before you come in and this and that. She would give me the name of the person who was on the phone, you know. This is so and so do you want to talk to him? Who the hell is that guy, I don't know who he is, well, okay. But now you just buzz through and I answer it myself. Even up here I sweep the floor and make coffee. Past summers, even when it was so hot I used to wear a tie everyday. This summer I kind of cut that out, but maybe in the fall I'll go back to a tie.

The best thing, there are a couple of positive things, I would say. Number one is that I think without my effort the J&L Structural Steel Plant that is down there , would not be there, which was a 14 Inch Mill, if you remember. Now it is a separate company, I think without my efforts there wouldn't be those 170 jobs. That's one of the highlights that, together with Charlie Laughlin, who was a representative at that time, who is deceased now . . . We got some money through the state. 250,000 dollars. We did a study and once we found the study, we found somebody who was willing to invest in that. We worked on it and put a contract in, made sure that the company who was coming in could make money. The guys could make money. The guys could make money and are working and they have been there since '87 now, so that's one good thing.

Also, second is like we've been able to keep this plant partially open here. I think that's through my efforts cause we were marginal. But I think 'cause of my efforts, and the union's efforts if you want to say that, that we have been able to make sure that there's still a plant here in Aliquippa. It's not like caving in neither, we did our thing. They're able to get early pensions for guys who were going to lose their jobs without it even being in the contract.

We got pensions for guys who really weren't entitled to them, because they didn't meet all the requirements of the contract. They were entitled to them, but you got to be off two years before you're eligible. These guys weren't off two years, they were like off a year. The company could've got away without them paying them and took them to court and you lose. But through my efforts I was able to do that, you got 385 guys' pensions. The company, when it was going through bankruptcy, they took the 400 dollar supplement and they took the insurance off the thing, got it back for the retirees though. Those things I think are my highlights, so to speak.

Yeah, the shutdown. You know, shutting down the Blast Furnace, the Coke Department, the Steelworks, seeing the buildings being torn down. The shutdown of the mill, I don't know how you can get any worse than that. The reaction in Aliquippa, it was terrible. It was just like something that was happening, like in slow motion, that's bad that you can't stop, but you see it happening and you know, that was the worst for sure.

People thought it would reopen for years, for years. They thought it was going to come back on-line. That's why when the company came to me and said, look, we're goin' to tear this down, they thought I would fight it, or this or that. But the longer that plant stayed idle the worse it got. And the new technology, they would have to tear out and put a new one in anyway.

The way I looked at it, the longer I compared it to a dead horse, the longer it was there, the more it was going to stink. The sooner you got it out of there the stink would go away, and that's why it's level today. You got to start with something. You still had those Blast Furnaces standing there, they would just be standing there, that's all it would be. It would be getting rusty. People would be driving by and saying geez, look at that, so that's the way I saw it. It was a dead horse, and the longer it stayed there, the more it was goin' to stink.

I don't know if it's goin' to be a gambling area or not. That's what they want, that's what certain individuals want, but if it is goin' to be that, it doesn't look like the state wants to. If it goes referendum it's not goin' to pass, you know that, the gambling is not a salvation for Aliquippa or Pennsylvania or anybody else. If they put gambling down there, they'll just drive through Aliquippa, spend their money, and then they'll drive back out. If you put a Ruth's Chris Steakhouse right over there, nobody will stop between here and there, nobody will stop. They'll just go gamble and go home. Put a gas station in, you think they'll stop for gas, it's just in-out, and my guess is if that is what we could get in, yeah, but I would rather see small industry.

My plan would be light industry and manufacturing, light manufacturing. Geez if we could get something like IBM, Intel, and Microsoft to start building little plants down there making chips, disks and that, have about ten plants in there, each one with 500 or 1,000 people making toasters and ovens . . . much better. Much better than gambling, but if gambling comes I wouldn't say no. I wouldn't vote against it or anything.

The former J&L site has been parceled out. CJ Betters owns part of it, and then I think that Beaver County Development owns some of it. He owns like from the tunnel north, where the Blast Furnace was that way. He owns them, and this way is owned by LTV and this Beaver County Development.

Paydays were memorable. It was like a riot, but without fighting. It was just like you couldn't walk down Franklin Avenue without shoulder to shoulder people on both sides of the street. You go up there and you used to get paid in cash and they had all kinds of guards up there with rifles and shotguns. Used to go up there and get your cash, you used to get two dollar bills and everything. Pay day was a big-time day, every other Friday.

Well, most of them come down the street and spend their money. Whatever they needed — if you wanted a hat, they had it, your clothes, you had shoe stores, you had the PM Moore Building Supplies. You see that big building over there, it was five stores, Pittsburgh Mercantile, the best store in Beaver County. It was all good stuff in there, you could charge like to the company. It was a company store. When I went to buy something I bought it at a company store. We bought shoes, we bought hats, hunting stuff, everything, everything was there. But everything was good in that store, there was no junk in that store. Maybe late 60's or early 70's. They had a grocery store and everything there, whatever you wanted, they had it. That was Pittsburgh Mercantile. It's pretty hard to forget anything. I have no bad, bad recollections of the mill. To me, the mill and the union were good to me, and everything that you see that I have on I can attribute to the mill, to the union, and even to the company 'cause they gave me the job. I give the union the credit because they negotiated and got it for me.

Because of the union I have been able to build a house, bought many cars — all American made — sent three kids to college. One time some guy said I went somewhere on a trip and he said the union paid for that. I said, you're right, not only did the union pay for the trip, but they paid for them shoes, them socks, and these pants that I got on, so if you want to go back to where the money came from, it came from LTV through the unions effort to negotiate for me. So if you want to say them socks come from the union, yeah, that doesn't bother me, I can say that.

They said like I went on a Safari to Africa, I paid it with my money that I earned from LTV through the union that negotiated it with LTV. The union sent me on that trip, if you want to say it like that, like I got money under the table. But I say, yeah, the company sent me, the union sent me. Yeah, I attribute everything I got to the union.

Well, once J&L, the old company, was a good time honored company. But the merger, once they merged, when LTV took over J&L, that was the beginning of the end. Because John Ling wasn't a steel person, but once he left, once he got it and that was it, and then they merged with Republic Steel, it's been downhill ever since. Went bankrupt. Now they are back on the even keel, it looks like they're going to make it. But we'll see what happens, that's all, but I think they're okay now. So they have opened up, they are basically one plant here, a Tin Mill.

Well, to me I've been lucky. I work, so it was really no effect for me. I think the shutdown could have been prevented. Imports number one, over capacity of letting subsidized steel come into the country was bad. It was tough enough to compete, but when subsidized steel comes in and subsidized cars, you can't compete against subsidized goods. We were free enterprise and then those other plants were bringing steel over and that was particularly paid for by the government. You just couldn't do it and they'll take a hit on there just to keep their people working, so . . .

The mill, it's still running, we don't have no last day. That's right, part of it, but the Tin Mill, nothing else. I wish that you would've been around to see the way Aliquippa really was at one time, and the way the industry was and what big steel was in Aliquippa, and that's probably about it.

No, I don't have any regrets, I can't really blame the company. I can blame them, but there's enough blame to go around for everybody — the union, the company, the government. But I don't think anybody really foresaw all this. If somebody did, maybe it could've been changed but it was a preventable, some, yeah.

Well, once LTV bought 'em out there was no choice on their part. That's just like, what's his name, Millkin, you know what I'm talkin' about. Yeah, corporate takeover, LTV in 1977, and Republic Steel in 1983. Once they do that, they don't think about people. Like I think once the steel people that were still running it, they worried about the steel industry and you got to rebuild this and you got do this. They knew the steps, but when you get a guy like LTV come in and then once the merger comes in, then you had too many people that were never tied to Aliquippa or whatever.

Strike food bank

Strike food bank

Nick Mamula

Union officers

Union officers

President Lyndon Baines Johnson and George Meaney

President Gerald Ford

The American flag

Chapter 25

୧ୠଔ

Management

H e was to retire that week from his foreman's position, an uneasy peace between labor and management, never quite part of either world, but an essential part of the day-to-day operation.

My name is John Cindrich and I'm from Aliquippa, PA. I'm sixty-two years old and was born in Aliquippa, PA. I had forty-four years in on February 11th of this year and was hired on February 11, 1952. I think I was making like sixty-five cents an hour. I'm on salary right now, I'm making almost 4,000 dollars a month.

My first supervisory position was February 1, 1968, I was a Welding and Burning foreman at the LTV Plant. I was assistant Welding and Burning foreman. Well, at that time, there was everything there in one place. That mill was like self-contained. We did everything internally, so in our shop we built all the fabrications throughout the whole mill. So consequently we had eighty-six welders in that department that I was in, and boilermakers, which was a fabricating shop. We were certified welders and we were like the boilermakers, a self-contained contracting company. So we did that work, the riggers did their work, the blacksmith shop did their work, the machine shop did their work. I was part of General Maintenance, right.

John Cindrich

The fore-
man's job was to
assign men to
work with differ-
ent crafts and any
problems that
arose throughout
the whole mill.
Whether we
were going to do
the job, or if the
department was
going to do the
job. We went by
step to our expert
teacher, it was
called, to help
our people do the
job. If not, then
we would bring
our people to do
the job for us.

Well, a typi-
cal day, the
hourly guys
started at seven
a.m. and worked until three-thirty — eight and a half hours, half a hour
off for lunch. I would have to get there at least an hour before, and I'd
have to make up the schedules to put out for the turn foreman and give
them drawings and get the men assigned their work. As the job changed,
I'd have to go out with them ahead of time, scout the job, go over the
prints with them, and stuff like that. The turn foreman worked under us,
so to speak, 'cause we worked around the clock, twenty-one turns a
week, 365 days a years. So there were turn foremen and they did the
same thing we did, but on the off shift. But I'd have to line up their
work.

That's right, I started off hourly, in fact I started off I was Dues
Protest Committee, when Mamula ran for office. I helped him get into
office, and then from there I was transferred over to supervision. I don't

think it changed that much with me because I was supposed to lose my job in 1985. The hourly guys had a big blowup, which was unusual. It wasn't unheard of but . . . The way it turned out, I was offered another position, so consequently tomorrow is my last day full-time — ten years later, so.

My good day, when things seem to be going right, but there was hardly things that seem to be going right. Something would always disrupt it. My best day is when I was able to go to other people, and they were able to help me or I was able to help them.

Management response is that the workers are the problem, that might be true to a certain extent, but whenever that mill was in trouble the hourly guys always responded. They would bust their hump to get that mill back in operation. I seen that time and time again, Blast Furnace rebuilds or breaks down or anything else you want, when those backs were to the wall, those guys would produce time and time again. When things were at a lull or normal operations. Granted, they did take long breaks.

The worst thing that I remember is when I was supposed to retire and I didn't. They offered me another job, and I had a crew of men and we had to go in the mill, and they were selling off the assets. We would just go in there and literally just rip things up and put it on the truck and sell it. Things that for years I maintained. I was just literally ripping out. Selling it for scrap and selling it for parts. To me that was the worst.

No matter what side of the fence I was on, I was always on the hourly side or the salary side and there is a good group of guys over here and there's a good group of guys over here. I think that's what helped that mill gel. I was fortunate enough to be on both sides. I think worker-management relations changed, but I don't think we realized the change until after it occurred. There was a difference, and we can look back and say it was better under J&L. Because J&L was more community related, they did things for the town, they did things for their employees.

When it went to LTV, that started to switch the other way. More so even later on when there was really in financial trouble, where the other guys were asked to make sacrifices and the salary guys were making sacrifices. Even in the end, though, the hourly guys got their sacrifices back and the salary guys didn't get all their sacrifices back. Like right now, salary guys still only get three weeks vacation and the hourly guys got all their vacations back. Things of that nature.

There is a colleague I remember that just passed away. We worked together for forty-four years, just passed away on February 13th. I often thought about that and my final response is that I don't think I could've changed anything, and I don't think anything should've been change. I think that's just the normal way things are up here, I don't care where you go. Things are the same no matter where you go. I was fortunate enough to have a brother who worked in the mill and he left and he went through different phases, and I told him, how can you go from making steel, to making plastic milk bottles, to makin' milk, to makin' prosthesis? He told me, he said, no matter where you go your problems are still the same.

Young people today, I'm glad I'm not young in a way. Too many young people today. There is no loyalty on the company side or on the employee side, and the way conditions are in the world and in this country today . . . I can't to say who's to blame or what it is, but I can't see to many people have forty-four years in one type of employment.

I'll tell you, I was just put on salary just for a few years before that, and I grew up with your dad, my brother grew up with your dad. That morning when I heard your dad was a fatality in the mill, it really struck me. I heard of fatalities before but nothing.. but never this intimately. Like I said, we grew up on Baker Street together, your uncles and everybody else, and that really struck me that something like that could really happen.

He was working on a crane in the Strip Mill. As I remember, there is a walkway and he was, like, out on a rail when he burned himself off there. He fell down and on the way down there was a gantry crane and his head hit that rail. It was about thirty-five or forty feet down.

I could never figure out, 'cause your father was conscious about that. He was overly conscious about that. He wasn't one to go in and get it done. He looked back. And I can never understand how that happened. I talked to my brother about it 'cause, like I said, he was a foreman of the riggers at that time. He couldn't understand it either, and I don't know, I could never understand it.

*****

It had been fifteen years since I had stood within those walls. These were the coveted jobs — the South Mill, clean, the Finishing Mills as opposed to the North Mill fire and smoke production. He seemed a quiet, understated man speaking from an austere office above the din of the noisy Tin Line floor — as he related more business and economy and less day-to-day personal drama.

My name is Bill Stephans and I live in Moon Township. I am forty-eight years old and was born in Pittsburgh. I have twenty-six years of service after being hired in 1970, April.

I was hired as a metallurgical engineering trainee. The training program consisted in metallurgical engineering was a little different than your regular training program. Metallurgical engineering allowed us to move around to different departments. We generally spent two months, two to three months in each department of steel making, finishing. At that time we had a lot of different ones, Wire Mill, Tin Mill. I spent time in the steel making — Wire Mill, Tin Mill — probably a little bit of time in the Hot Strip Mill, Seamless Pipe Mill, Continuous Mill, Welding Mill. Basically it was kind of a rotation for one year. Then, after one year you're generally placed in some kind of assignment.

I think there were about ten or twelve in my training class, now, they weren't all engineering people. Those were maybe a couple engineers, there were three metallurgical engineers. The goal of the program then was to get people from that kind of background into management, into the metallurgical quality control area at that time.

Outside of quality control organization, my first position was, it was called metallurgical investigator. What it was is mainly involving quality control functions, investigating whether . . . for example is it in the finishing part you might be investigating a customer claim, doing a history, doing background, doing metallurgical analysis. In the steel making end it might be determining heat analysis by determining what kind of quality control parameters were included.

Basically I was in the metallurgical quality control department for four years. Then I came to the Tin Mills in Operations Management. Operations basically consisted of responsibility for some particular operating unit. In those days I had a small area of responsibility for example in the Tin Mill. Now, I'm responsible for everything production — cost, yields, labor relations, and safety. Basically that is what an operations person does, he has responsibility for the operation of a unit, or a group of units, or a plant, and has the responsibility for everything that goes along with that — environmental issues, whatever.

Same thing only on a bigger scale now. Right now, I have my position and my title is Manager of Operations. Basically, I'm the front manager, I'm the only operations guy here. I have responsibility for

everything in the plant — environmental, safety, the maintenance of
cost, profit and loss (P&L). Then I'm responsible for everything, and
then reporting to a general manager and vice president.

Well, to give you an idea, in the Tin Mills product division we are
division-alized. There is a flat roll division, welding division, flat roll
plants, and the major plants, Indiana Harbor and Cleveland. There is
also a Tin Mill products division which is what I am part of. We have
another plant in Indiana Harbor that basically does the same thing. It's
within an integrated steel plant and we report to a vice president. There's
also a tube division. Basically, they are set up so there are sales and
marketing and operations and responsibility within the division. Levels
have been cut out. When I came here you had the plant manager who
had the responsibility for the whole plant and he had a couple of assis-
tants. So between the turn supervisor on the floor to the president of the
company there was probably twelve layers. Four layers now. There's
turn supervisor, there is another level between him and I, and then there's
me. Then vice president of the corporation. Basically it's flat, many
less people.

I guess one of better experiences — there have been a lot of good
ones — but probably one of the better experiences is the day that we got
approval to spend about fifteen million dollars. That's public record on
a new annealing facility. That was the first time. That goes back to the
late 80's and that was the first time that we had spent anywhere near that
kind of money on this operation, and took us quite a while to get there.
It probably took us three years to solve that job. What that did was,
number one, it showed the corporation that this place was going to continue to operate and it was a good day. It kind of made a lot of the things
that we did up to then worth while.

The improvements that we have made were in terms of costs, yields,
and efficiencies. What we have to do now is for customers. Now there
are changes in attitudes about customers now versus what they were
twenty years ago. I mean, the customer was just out there and we just
did our jobs everyday and the steel went out the door. After that now if
we don't talk about any particular customer or something that we must
do for a customer ten times a day, it's an unusual day. That's probably
the biggest difference, our reactivity to the customer today versus the
way it was twenty years ago.

What happened I think it opened a lot peoples eyes to what could
happen. I think it was a negative experience to a great extent, but I think

Ben Morell

it also brought a lot of people together who said, we aren't going to let this happen. I think everybody, from the people on the floor throughout the whole organization, said that if we could do anything to stop this from happening we would do it. I think, for example, around the same time when the rest of the plant was being shut down, we were going through some remanning. We were reducing jobs, hourly jobs, that kind of thing, and it went very well. We sat down with the union, we worked all these issues out with the union ahead of time and tried to get to what we thought were some of the more . . . the most efficient ways to do things, manning wise, and it went well. It went well because we communicated well up front. We did it by sitting down with the guys ahead of time and talking about different ways to do things. It was the first time we did that, we've done it twice since then.

We went, we never felt controlled because in every subsequent time we did it, there were always a lot of discussions upfront and a lot of issues settled upfront. It went well, basically because we settled all the issues ahead of time rather than go and do it, and let the issues come out, which was done in a lot of other places. We did them upfront. A lot of cooperation from the union, a lot of cooperation from the people, and I think part of that leads back to seeing what can happen.

BF Jones III

The worst day that I ever spent, I'm afraid to say, was on a fatality. That was the worst day of my life. I have had my share of bad days, but none like that. That sure puts it into perspective. It's very tough. I guess . . . like I said I'm forty-eight years old and I have two kids. I have a little more flexibility then some of the guys that work with me in terms of hours and that kind of thing. I like to spend a lot of hours there. But on the other hand, some of the things I had to do when I was first married, for example, I don't necessarily have to do. I have a little more flexibility.

Like I said, I want to go get my wife a birthday cake and take some things cause my kids go to bed early, and most nights I get home at seven p.m. I get to see very little of them during the week. Special occasions I try to make a point to be there. It is tough on the family, especially my wife.

Capacity hasn't changed much, but our efficiency is improved. Our cost of production is down. I think you can do awful lot of things if you can get everybody working in the same direction. It takes a lot of communication and a lot of hard work and a lot up-front explaining and working toward things and trying to make people understand. You have

The superintendent

to have the right mix of impatience to get things done with patience, to know that people aren't going to jump and say, yeah, that is the right thing to do, let's do it.

I think the best piece of advice I can give somebody in a manufacturing environment like this, you got to get the people on your side. You may use different ways to do that, but you got to get everybody working in the right direction. If you don't and you just try to dictate or whatever you might be successful for a short haul, but not for the long haul.

Management

Management safety rally

Management inspects job site

Management in job site inspection

The management office – back in the day

The management office – post shutdown

Chapter 26

ℰꙬ

Security

He must of been an imposing man in his day, tall, substantial, but soft spoken with a sense of fairness about him yet slowed by age and the unrelenting passage of time.

My name is John T. Sudar and I live in Aliquippa, PA 15001. I'll be sixty-seven years old next month, on the 10th of next month. I was born in Monaca.

I worked for thirty-eight years and nine months. I was hired in the Tin Plate Department in 1948 and was making fifty-seven cents an hour. When I retired, I retired in 1985 as a sergeant, I was in the supervision, I was making about 2,500 dollars a month.

When I first got hired there, we were known as what they called J&L Police Department. That was a spin-off of the old coal and iron days. You probably hear so many people talk about the old Coal and Iron police. We had a hell of a time trying to live that down.

The Coal and Iron police, from what I can gather — like I said, we had a hell of time living that reputation down — were strike breakers and stuff like that. Have you ever gone to the library and gotten hold of a book called *Boot Straps* by H.G. Mark? Get a hold of that book. I have never had an opportunity to get down to the library to get a hold of it.

John Sudar

But the guy's name is H.G. Mark, he used to be the director of the police, the big shot, the captain up at Pittsburgh City Office.

My time, it was the tail end of those coal and iron days. It was rough but we prided ourselves on being efficient and it was like a little city in itself. The mill itself was like a little city, it was all mapped out into roads. We had route numbers and everything, we had the north mills, we had the Welded Tube, and the south mills. We had to know a little bit of everything that sort of transpired during the operation as we would go into the department. We'd be able to spot it, if anything was out of the ordinary. We were able to notice if something was unusual and what have you.

At that time, when I first got hired, this is just before the time clock was going to come in. They hired a dozen new men, twelve new men, and out of that twelve men, there was twelve of us that was supposed to be hired for the time clocks. I got in just under the wire before the time clocks. As a matter of fact, prior to that we use to have only two men that worked a turn. There was a sergeant, a Number one man and a Number two man. The Number one man handled everything up around headquarters and the Number two man handled traffic and things like that. We took our turns about working the gate and took our turns about going out on patrol.

We had to investigate complaints, we had requests, and we had our share of thefts. We had to investigate thefts. If an employee would call in and make a complaint, we had to go and question him, talk to him, the supervisor, and investigate the scene and try to determine who committed the theft or what have you.

There was Headquarters Gate, there was the Brewery Gate, Byproduct Gate, Welded Tube Gate, A&S Railroad Gate, South Mill Overhead Bridge Gate, and South Heights Gate, plus the Slag Disposal Gate — so eight different gates. Oh, yeah, not only one, we would have three guys at headquarters. One guy at each other gate per turn, two guys at the Tube Mill Gate, and one guy at each other gate. So there was a total of sixty-seven men.

Well we had to cover everything at all times, toward the end there, we had when we went to work, depending on what turn it was. If it was the midnight to eight turn, if we went in on the midnight to eight turn, we had to make out the time. We had to tend to anything that came up like the normal chores, which would've been the traffic and the change of turns, and we had to go out the mill on patrol, and we also patrolled the outside perimeters of the property. Like, say, Paul DeVaney would go on vacation, we had to go and check his house, make sure everything was all right. Not only Paul DeVaney, Harry Saxler would go on vacation. We had to check his house on the four to midnight and midnight to eight turns we checked their homes.

Their houses, like Paul DeVaney house is up in Plan 6, Saxler's home was over here on Broadhead Road, you know, where Presbyterian Church was. Then we had to go up to . . . we went as far as Moon Township to check homes of some of them big shots went on vacation. That was four to midnight turn and midnight to eight turn.

There was a lot of thefts that were going on. They had big cases down at the Beaver County Courthouse, and not only Beaver County but Allegheny County, too, for the theft of molybdenum. That's an ingredient used in steel making. These people were taking mill supplies home.

They were doing on a wholesale basis, they had a truck. They had an inside man, he would gather up a lot of stuff and would have a dump truck come in. They would load that stuff up on the truck and would come in through the side disposal gate. They would come in through there because years ago they filled in Crow Island. They had the Basic Oxygen converter there and that was where they were making these thefts, from the Basic Oxygen converter.

The best day that I ever had, they were all good. The time that I spent down there, believe it was enjoyable. The men that you worked with, you got along fine. We usually started to work with a group. We were always assigned to different crews. Say, like as a supervisor toward the tail end, when I became a supervisor, I worked with all the crews. We . . . when I was guard and I was Number one man, we worked with a certain group. We had a . . . I worked with one group, one crew I worked with for about eight or nine years. The left hand knew what the right hand was doing constantly without any problems at all.

To be honest with you I couldn't really say that I had a really bad day because you got along with everybody. The guys that would approach you, well, I myself I was an old logstowner down there. I don't care where you come from, there was no place like Logstown. You know, you used to have to live there to appreciate where you were coming from. Anybody could approach me and say, hey, John, I need this, I need that, fine, I can't buy this anywhere, I can't find this, okay fine. Don't steal it, you know, when you have something, if you are going to have it made, usually they had to have it made. Let me know when you have it, and when that would happen, I say okay and I would talk to my boss. The captain and I would tell him, hey, so-and-so needs this or needs that. The he would ask how good of a friend is he. Very good friend. Okay, I'll take of it.

Why the hell anybody would want to steal anything from the mill is beyond me. Because you could borrow anything in the whole wide world from the mill. I'll tell you a story, a guy . . . I was standing at the gates this one day and I'm looking. I see this guy walking up toward me, a friend of mine, I've know him for years. And all of a sudden I said, holy good God did this guy put on weight, I said. Pat Hallisey was working the gate with me. He was my Number three man, he said, let's knock this guy off. I said, no, let's wait, so the guy hobbled around me, and he went home.

The next day we were on four to midnight turn and this guy is coming out again, back to his normal self, you know, and I stopped the guy and said, you lost a hell of a lot of weight, did ya? His face turned every color in the rainbow and he said, I don't know what the hell possessed me. I said, tell me something, what were you stealing, what were you taking? He said, believe me, I realized that I could've borrowed anything that I wanted after the situation had arisen. And I said, all right, what was it that you took? Two hundred feet of electrical cord, he was building a house. He had wrapped it around him, two hundred feet of cord, imagine how big the guy must of looked like. So I said, what are you going to do with it? He said, I'm going to make extension cords.

So we had a guy, captain, his name was George Stauffer, one of the nicest guys you ever want to meet. But to this guy I said, let me tell you something, after you are through building this house I want you to let me know. I want you to bring that stuff back and we will take it down to your department and give it back. He said, I'll catch all kind of hell, he said, if my foreman finds out. Your foreman won't know a thing about it, I'll just return the stuff and we will take it through that way. He said, I'll be on his crap list, and I said, no you won't be on his bad list, I said, I'll take of it for you. So after the guy built his house and everything, he took the stuff down.

How many of those little incidents had occurred, you would never know. As a result, I used to be able to go into any department throughout the entire plant, not just the north mills area, but throughout the entire plant.

When I first started there my dad was a . . . he was a musician, he was an old timer. I was only twelve, thirteen years old when my dad died. So a lot of these older men didn't know who I was. So I walked, I was out of the mill on patrol, and I walked into this one particular department, it's called the Open Hearth lower level, foreman's office. All the old Serbian guys were down there and when I first walked in there, they said in Serbian, hey, my God look at this big shot walking in here. So I didn't pay any attention, so they were talking.

I said, okay, I didn't say anything, so after time went on they start cutting me up and calling me a biscuit snapper and J&L stool pigeon and Jesus God you wouldn't know what else. So when it came time for me to leave, back in those days, we only carried — we had a gun, we had a blackjack, and a nightstick that was issued to us. But back in those days we didn't have radios, so I picked up the telephone and I call the guy that was working the desk. I told him — I didn't say what my last name was — I just told him I'm down at the open hearth lower level and they were kind of strange that I didn't say my name, and he said okay, fine.

So I turned to the guy that was . . . you can always pick out the ring leader of these guys. He was cooking that day, they took turns about who was cook. It was on a four to midnight turn, so I turned to the guy and so I said in Serbian, well uncle, I'm leaving now. I didn't learn to talk any hunky till I started working down the mill, because my mother and my dad never spoke any Serbian-Croatian unless they didn't want us to know what they were talking about.

You always try to respect your elders by calling them uncle or aunt, you know, you remember that. So he said, oh my God, who are you? I said, you're going to see who I am. So with that I just let it roll off my back and I laughed so hard. So these guys used to see me down on the avenue and they try to . . . they made it their business to find out who I was, who was J&L stool pigeon walking in. They were going to find out who this guy is that can talk the hunky, and I could.

I could hardly speak that much, but anyhow, I got an education down there and they use to try and drag me into every tavern on Franklin Avenue. Franklin Avenue. At that time J&L paid off in cash and every second Friday was a pay day. Oh, let me tell you, we used to have guys come in, trying to get to work half bombed up and what have you, and we used to have to turn them away. We use to have to fill out reports. How many times we used to have to go in before labor relations and what have you, and Tony Vladovich was the president of Local 1211 at that time. There were only two hunkys in that whole J&L Police Department, it was me and Sam Muslin. Sam Muslin was a heck of a nice fellow.

But what happened, both of us were working that time and this guy wanted to file this . . . a guy that was turned away wanted to file a grievance. He wanted a day's pay and Tony Vladovich was reading the report and he said, who was the guard. He spoke a little broken English, but a very educated guy. He try to give you the impression that he wasn't well learned, but he was a very well educated man. Who was the guard, who were the policemen at the gate when this happened? John Sudar and Sam Muslin, and he threw his hands up in the air and he said that's all, get out of here, he said, you don't have a case. That was one of my best days, the union guys, they worked with us without any problems.

Like Melvin Kosanovich, Melvin was a heck of a nice guy. I went to high school with Melvin and he worked in the carpenter shop, and later on as time went on he became an officer in the union. I could always ask Melvin, hey, what is going on, what is happening here or there, and he use to tell me. Like I say, back in those days if you were from Logstown, no matter what . . . that's like your dad, I used to go to him and say, hey Matt, what's goin' on, or your uncle Lubar, what's goin' on in the rigger department. They would tell me whatever it was with no problem.

There was cooperation between the hourly workers and the police department. There were some guys that you couldn't get the time of day. You know, from the hourly men, it all depended on if you were liked. If you were liked, they would tell you anything or do anything for you.

There was not just your dad and your uncle, there were a lot of people that I knew that I grew up with, and they could come to me for almost anything, anything in the world, they would come to me and say hey. My nickname in high school was Knobby. They would say, hey, Knob, I need this, I need that. Don't worry about it, we'll take of it, when the time comes we'll take care of it and as a result we got along fine.

I used to work at what they called the North Mill Relief job. I would work two days at headquarters and one day at the Brewery Gate and one day at the Byproduct Gate and one or two days at the Tube Mill Gate. Like when I would be working the Byproduct Gate or the Brewery Gate, whichever way those guys would come in they would say, hey, come on down. I just snubbed them. I would say I can't make it tonight, I can't come down there. What's the matter? I said, I have to be here. We'll take care of you. They used to bring stuff to the gate for me to eat and it was something, you would think that something unusual.

Well you, Joe, you know Carol, and my youngest one, Andy, they were active in all different kind of sports. Between my wife and I, we were at about every ball game and we put the kids through college and they got good jobs now and everything is fine. My last day, it was very emotional. I had a hell of a time. That was shortly after I was off sick for a while, and I came back. I was in the hospital when this fellow called me from Indiana Harbor. Captain Morgan was his name, I broke him in originally, but he is a hell of a nice guy, too. He called me from Indiana Harbor and he said, hey . . . as a matter fact he called me right at the hospital. He said, what are your intentions, what are you going do? I said, I think I'm going to pull the plug, I think I'm going to call it quits. He said, now would be a good time to do it, he said, because you can get six month salary severance pay, and you can get six months unemployment compensation you can collect. In other words, I could get a year's pay without having to work.

So right after I talked to him on the phone I called down at the mill and I talked to a guy by the name of Ray Tremont. He was like acting director of the Security Department at that particular time. He happened to be down at the plant and he said, well, what do you think you want to do, John. Joe Rossi, I think his name was. Go over and see him, tell him. I knew that big shot when he was a little kid.

John Sudar on the job site

I said, you want some bodies, I can be one of them, and he said, oh. I said, you're planning on laying some people off anyhow. He said, yeah, but we aren't going to lay off any supervisors and you are a good supervisor. We don't want to lay you off. Oh my God, if Bill Milne — Bill was also a sergeant — if he finds out that you are going to go, he said, I know he's going to want to go, too. I said, so what, I said, that is two bodies.

So he went over to the main office and he came back. He called back about twenty minutes half hour later, I was still at the hospital and he said you can't retire while you're off sick. I said, well how about I work a month. He said, fine, he said, he'll agree to let you go. I said, good. He said, not only that, he said, Bill Milne grabbed me just as soon as he found out that you were going to go and he said he wants to go, too, so we both retired about the same time. My last week I had a heck of a time. All I was doing was going around saying good-bye to everybody.

I would tell them that we had a lot more camaraderie then anybody could think about transpired and went on down there. The working man and the supervision got along. I see some of this stuff going on like with US Air, that's following a trend just like the steel workers. They are going to work themselves right out of a contract, right out of a job. Show me anybody that is worth fourteen or eighteen dollars an hour. There isn't a guy, maybe in the medical profession are worth that kind of money. But you didn't get paid for your brains, it's what you know. You got to practice.

The shooting range

Chapter 27

ℰℭ

Health Center

S he entered the steel profession as a new graduate nurse, cap and
shawl, looking like the proverbial Florence Nightingale. Early, she
labored on the wounds and injuries from the steel making process, while
later there were tales of dependency and depression with the lack of
work.

My name is Janice A. Paul, RN and I'm from Aliquippa, PA. I'll
soon be sixty and was born in Monaca. I have thirty-seven years of
service and was hired in 1959, March and I was making probably about
4,000 dollars a year, about 400 dollars a month.

Back in the 50's and 60's, well, we had our own little health center
and, at that time, we had our own lab, our own x-ray department, our
own audiologic department. We did a lot of emergency treatment. We
were basically like the hub. There were three other dispensaries, we had
two dispensaries that were open twenty-four hours-a-day. The main one
was the one that we had our own little emergency center and we had a
clinic there. We had an orthopedic specialist that came in two times a
week to see our patients, and we had an optometrist that came in every
day to see our eye cases. We had a staff, we did all our own physicals.
We had a staff of two full-time doctors and two part-time doctors, and
we had a crew of sixteen nurses and, like I said, we were open twenty-
four hours-a-day.

Janice Paul

We had our own physical therapy department and we did . . . we gave whirlpool treatment, the ultrasound, the electric muscle stimulator, the whirlpool, traction exercise and we did the range of motion exercises for our back people so we had full function. On an average day we probably would run just on daylight. We had sometime five nurses. We ran on the full twenty four hours, probably about 150 patients we ran through there. It was quite busy.

At least 150 patients a day, but that's all three turns. We had some really bad, bad emergency cases. There were a lot of amputations, crush injuries. We had several deaths in the mill at that time. Well, we also treated the outside contractors. Yeah, we treated them and if they were real bad, we would stabilize them as best as we could and then we would send them to the hospital.

It was Aliquippa Hospital, it was always Aliquippa Hospital in a real bad situation. We would treat the industrial cases, too, but Aliquippa was always our focal point after they became stabilized. If it was a nonindustrial worker case, they were permitted to go where ever they wanted, but Aliquippa was always our hospital.

Well, I would like to believe and like to say that I think that they got top quality care. I think that our staff, the nurses that we had at the

New nursing graduate Janice Paul

time . . . we had a really good staff of nurses and we have always had good communication with Aliquippa Hospital. The girls have been very receptive to giving us information and many times we could go up there as often as we want. Our doctors at that time . . . my old boss was on the staff of Aliquippa Hospital and we would go up once a week and he would take me with him and we would make rounds. We also did the death calls. If we had a fatality then it would be the responsibility of the nurse and the doctor to go to the home to help the family to do whatever. We would go into the scene of real traumatic kind of things. There is a lot of history behind here. We had 15,000 employees then.

So, if there was sort of an industrial accident, you would go to the home of the family. We would be there, we would be the first one there to tell them, and offer emotional support. Right, and do whatever was necessary, wait until a relative came. We would call the undertaker for them, would call the coroner. We would take care of all their details that were necessary until someone got there to help them, but it's different now. I'm just kind of like a band-aid nurse, now. I take care of the kind of compensation things in the building and, of course, we have a lot less employees, too.

My best day is when I'm taking care of people, the patients. I know most of these people because I'm a Valley person and I've lived in this Valley all my life. But most of these men I know so they aren't just my patients, they're my friends. So my best day is when I'm helping them. I enjoy that, that's my life. When I don't have any patients, that's my worst day.

On the average the most I see is about ten patients, and its nothing real traumatic, nothing that is earthshaking. The worst thing that can happen to, you know. It's because I no longer have any equipment except First-Aid, to really try to stabilize anybody. I have a small oxygen tank, but if someone comes in that is having chest pain, then I got to immediately figure out what is wrong. I must always send them out 'cause I don't want to take that responsibility of missing anything.

Well, there are some real bad things that stuck in my mind, and they were the traumatic injuries that we had. It's kind of the on scene things you do, you know to do. Well, there was a time when we had an outside case where it was a young individual, he was working for what they called a supersucker and he got pulled into the supersucker. So he was . . . his body was completely dismantled, and myself and the doctor had to go down to the scene and it was very emotional for me. I really reacted to it nights later, because we were going down to pick up body parts and this was a young individual. It didn't matter if he was young or old, it was just the fact that he was a human being and it was really hard for me to deal with that later on. I came to grips with it but it was difficult.

There were other incidents of traumatic things that I had to go into the mill. We had some people that fell into the river and I had to go down onto the barges. We had another man that was buried alive and I had to go down there. Those were traumatic, but not as much so as this one particular incident. I remember that, I will probably remember that all my life. Industrial life, it was difficult for the men. They work hard, I believe they work hard at what they do.

Yeah, my one boss, she was a great person. Her name was Grace Lang. She was a special person to the men. They just loved her because she was always there to help them. I never saw a mean side of this woman. She was always the same, and now I believe she's probably in a personal care home somewhere, but she was always my idol.

Industrial nursing, it really didn't have any affect on them. I raised three girls and my one daughter is a nurse, so I must've did something right. She's working in the emergency room and she loves her profession.

I was able to help her through the rough parts of the deciding factor of I-think-I'm-going-to-quit, and all this kind of stuff. But she can see all the rewards that's there now. Not money rewards, but the rewards of treating the people because when — I believe that when you see, you can see the joy of the people, they wait for you. They say . . . not that you need a pat on the back, that's not it at all. It's just the fact that you are able to give yourself to someone and you get back when you see them well.

My daughter is like that so, you know, I think that my industrial nursing has been something of a helpful hand to them also. Plus the fact that I was able to work three turns, which was good for me. 'Cause I could still take care of my own family and they never had to be without so I think that the whole realm of industrial nursing was good for me.

There have been times in my life, I worked in the operating room for two years before I came here, and I miss that. There have been times in my life that I wished that I would've stayed at the hospital. Especially now that I am just a First-Aid person, cause I miss the hospital. I went back in 1988 and took a refresher course and recognized how things have changed.

No, I've never been laid off, when I first started in 1959, they had the strike. I was gone for a while during that strike because I was a new employee, but I've never been laid off. It's disastrous, there were men out there looking for jobs and some of them were divorced from their wives, some of them committed suicide. Someone would come back and visit you and tell you that they had three or four jobs to try to maintain what they were getting at one job here, and they weren't making it. It was awful, they moved out of the area and then they would come back. Many of them lost much. We were lucky, those of us that remain, some of them don't recognize how lucky they are.

Right, and some of them have the attitude and I like to believe that the company doesn't owe anybody anything. No one ever asked me to come here to get a job, and I think the company has been good to me. I think the company is good to most of their employees, but you have a handful of people who like the company. I think twenty percent of your force likes to complain, but after a period of time it's just, like, it goes over your head, you just listen and don't say anything.

As far as change goes, I think the older men, they had all of their lives. When you have twenty years of service, you became an old employee. Well, twenty years of service doesn't guarantee you anything now. I think that during the period of time that the men were looking forward to getting some more seniority in their life, they thought they were going to have it easier. But it didn't work out that way. As they got older, their job became harder instead of easier because there was reduction of forces which the company had to do to balance the books. I mean, they were all doing the best they could for our own survival. It was just a change, a change with the whole society. Not just here, everywhere.

I would tell younger people that they need to get an education, to go to college, because they can't find a job at the college level and that they got to hang in there, eventually something will come up good for them.

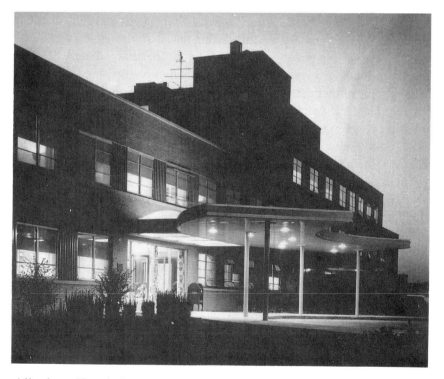

Aliquippa Hospital

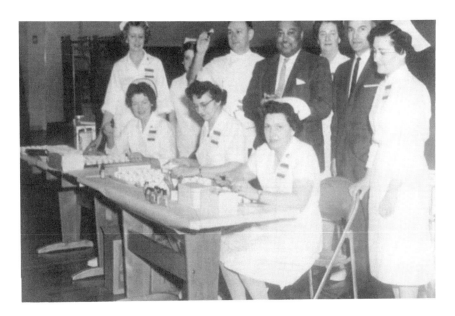

Workers health

Polio immunization

Chapter 28

ℰℭ

Railroad

H e looked quite distinguished, dark complexion and features contrasted with snow white hair and goatee neatly trimmed. The story was from the other side of the tracks, so to speak, working as a minority in the railroad trade. His thoughts were quietly stated, yet forcefully portrayed of working successfully in another man's world.

My name is George Stokes, and I'm seventy-five. I was born in Maryland, Hagerstown, Maryland.

I was hired at A&S in 1949 and had thirty-three years of service. I retired about 1981 or '82 something like that. Oh boy, that was pitiful, we made, I don't know exactly what we were making, but I know we didn't get paid for overtime. We worked six days a week. We made an hourly wage and didn't get paid for overtime.

Well, they were getting around to it 'cause the union, our union, wasn't that strong, but it started building at that point. And the next year we started getting paid for Saturday. They started paying us time and a half for Saturday, the next year, in 1950.

My family moved here. My mother and father are from Braddock, but I had an aunt over here in Aliquippa and came over here and stayed with her. Then I got a job in the mill. I worked in the mill first and then I came to A&S Railroad. Oh, our job was fantastic, we laid track . . .

George Stokes

Aliquippa and Southern hauled everything that was made in J&L to the main line, which was Pennsylvania and Lake Erie (P&LE) by the A&S. The Aliquippa and Southern Railroad took finished product, and hauled it to the Pennsylvania and Lake Erie Railroad system. It was the main line to get things movin' either east or west, wherever they were goin' but whatever came in had to come from PLE, too. They would bring it so far to our line, then our engines would pick it up.

Well, I started out as a track man, ended up foreman. The track man maintained the tracks, kept them repaired, put the bolts in, kept the ballast up good and tight, kept the railroad so it would operate. The ballast kept the road solid, kept it so the ties would sit on something solid instead of bouncing. It would sit solid, like a good paved road.

It was mostly hand, but then later we started gettin' tools and we started gettin' machines. Then the machines start doin' the work. But when I first got hired we did it by hand. We did it with shovels and picks, but later on we started gettin' machines and machines would tamp the ties. We used to do it with our hands and a shovel. The machines started doin' it with their things and they made it much tighter. They did a much better job then we could've ever think of doin'. Well, when we did it by hand we had ten and twelve people in the gang, and then they reduced it down to six. When I retired we were down to four in a gang.

Well, when I got to be a foreman, we laid track, the BOF what they built out there . . . this is what I never understood about this mill closing. The BOF that they built was a new mill, we laid new track. It was new land, they took land out of the river to build it, they put slag in there, and it was all solid land. They brought in oxygen from West Virginia to take care of the BOF. It was a brand new mill when they tore it down, it wasn't ten years old. This is why I never understood how the mill could be, 'cause evidently they didn't know about leavin' either. They wouldn't put that much money into a brand new operation.

And took it completely down. We laid the track, brand new track for the whole thing for the engine, the turnaround, everything. 'Cause it was a new section of the mill where they built this BOF and that was our job to. They wanted to hire it out to a construction company and our superintendent said we can do it ourselves, we can build it. We had guys who could do it. We had engineers, too, and we did build it.

Oh, I liked workin', enjoyed my job. I enjoyed bein' foreman there. It was, I guess, I enjoyed it because they told me you will be a track man as long as you stay here. But we got guys that come in and they could understand when people could do things. I could do my job, could do it real good. I was proficient in my job and knew I could do it good. But my job was good everyday. It was a damn good job.

Oh, the bad days was when it rained, cold and snowin' and you had to get the job done. You had to get it done, you had to stay out there in that cold or in the rain, whatever.

Oh sure, I remember a cold day, a wreck was up in our circle. They call you out for the wreck — you didn't have to come, but I came all the time. I always came and got a crew. I got two crews out to fix the wreck because it was a bad wreck. It was part of the mill that had to be repaired and they needed it to move the steel. The whole thing consisted of movin' that steel, keepin' it movin', so it was freezin', we drove the spikes in, and the freeze would kick 'em back out. It was that cold at night.

It was bad night and we worked the whole night. We worked the whole night through. Mostly we can do our jobs and get finished, go in maybe two, three, four o'clock, something like that, and they let us stay and they would pay us for it. But that night we worked the complete night through. That was a bad night 'cause we had to work, and we had to do hard physical work in horrible conditions.

I don't know, it was a nice job, it paid good. In the end we were makin' good money. I'll tell you the truth, when it come time to retire, I didn't want to retire. But then the guys that wanted my job, they said George, you workin' for fifty cents an hour. Figure out, don't count your overtime and stuff like that. You workin' for fifty cents and hour. So they convinced me cause we didn't have to retire.

No, they didn't ask me to retire. I didn't have to, but you know how you end up talkin' and the guys say George, you workin' for nothin'. See, you got to buy lunch, you got to buy the work clothes, you got to do this. Said count it up and your workin' for fifty cents an hour, take your pension. 'Cause I didn't want to take the pension, I had a nice job.

Oh yeah, the guys, well, we had guys that . . . the weather like it is today, this was our money weather, we made money in this, in the snow.

'Cause we had to keep the switches clean. We had to clean out the switches, the switches had to be clean so they could throw 'em and keep the steel movin'. They had to have the switches open and we finally got around to where we burned them open. They didn't like for us to do it 'cause it wasn't safe at all, but we burned them open. Then we finally got a real snow burner to take care of 'em, but before we got the snow burner, we burned 'em open ourself. But then they got a snow burner to burn them open, and that's how we kept the switches open, and those were our wintertime jobs.

Well, the tracks, we didn't care about, it didn't bother. We had to keep the switches open. The switches is what was important, 'cause the switches had to open up to go in and come back out. If the switches didn't work, the train would wreck . . . and if the train wrecked you in trouble then.

Well, at first it was dirty, I couldn't bring my clothes home. I had to take them someplace else and get 'em washed. But after I got to be a foreman I could bring them home 'cause I didn't get that dirty. 'Cause I wasn't allowed to work then, but all the foreman did a little bit of work because they were used to it. You know, and it was something you did physical. It was part of your exercise like. You weren't supposed to, but all of us grabbed a hammer and put some spikes in once and awhile.

Oh, back when I first got hired, oh, it was then, it was terrible. They told me you'd be a . . . I asked them right off, will I ever be able to move up on the train crew. They said no you were hired as a track man and you'll be a track man when you leave. He could tell me that right off, but I had to accept that because I wanted the job. I accepted it, but I knew I had a high school education. I wanted something else, too. I could do my math, took the test like everybody else. I passed all the tests that they gave us down here. I said I want more and got more, but not from that guy. Not from him. He told me right off, you will be what we hired you for when you leave here.

Oh, yeah, it lessen' up over time. Because they finally got black guys on the train crews and everything, but they started out there were no blacks on the train crew. The train crew was all white kids. No black kids worked on the train crew but there was black, they were stokers. They pulled out the steam engines when they came in to be changed over. They pulled the coal and stuff out. No blacks were to be around the engines when it was steam.

Yeah, when they switched to diesel, some blacks moved over to brake men and firemen. In the locomotive, two brake men, an engineer and a fireman, and whoever was head of the crew. It was five people. The engineer drove the train, he was in charge of taking the signals from the brake men. The brake men went outside the train, yep, they went out through switches, lined the train up, hooked the cars up to it, whatever it had to pull.

The fireman, he just sat. But when it was steam engines, he worked, he did the most work. But after they got the diesel, he was just there. He put the coal in, that was his job, but they were white kids. There were no blacks. In the south the blacks did that.

And why did that occur? Because of our society — it's racist, the whole society is racist — it's easy. Sure. It's proven that our society is very racist. The whole mill was like that, they had certain bathrooms that blacks wasn't allowed in. They had big parties that blacks wasn't allowed. That was 1949, so even after the war this occurred. The Tin Mill had no blacks, the Carpenter Shop, none of the trade units had any blacks. Electricians, none of the crafts. We got the labor jobs. Like he told me, you'll be a laborer when you leave. It was systematic. There were blacks hired into the 14 Inch Mill and Coke Plant where the dirt was. Jobs were dirty and heavy jobs. The Coke Plant wasn't heavy, it was just hot, dirty, but on the A&S between me and the next guy gettin' hired it was fifteen years. 'Cause they didn't mean to hire me. I guess they figured after they got me they got the wrong one, because I stuck up for myself. When we got a union I said, I know why I can't be promoted.

Why can't I do this, and if we got a union I'm goin' to pay you dues I want money for my dues.

Now? geez, it's all gone now. But they did change it at the end when they got to close the mill. The black guys were movin' around — I seen electricians, carpenters. They moved into every place, but when I was young, they went nowhere, they were in no crafts. Separate bathrooms in the 14 Inch Mill, in the Coke Plant, they had separate bathrooms. They had some people in the Tube Mill. They didn't want you there, they didn't want you to come through there. Well, it stopped when they started hiring all over, but it took a long time. Well, I'll tell you how it went in the 14 Inch Mill, they got to the place where blacks were makin' big money. When they started makin' the big money, the whites come in. They come to take those jobs. They used to call it the slaughter pen because they pile the steel up real high and it would fall, and the cranes would run over it, take it. It would fall and it would fall on people. They piled the pipe up and you had to walk between the pipe, and they had cranes stackin' it but the support wasn't strong enough and they would fall. When it fell, you're dead.

That department started shiftin' from black to white mostly as the wages went up so high. They were makin' so much bonus they would fire the guy with their time, man, whoever kept their time and evaluated him for bonus. He gave him too much bonus and they fired him. He gave the blacks too much bonus and they couldn't take it back, they couldn't switch it back. He gave the blacks too much bonus. That bonus was real high, and then everybody wanted them jobs.

My last day at work, they gave me a little send off and everything, and I made a little speech. It was good. I didn't want to leave work. I enjoyed it, had a good job, paid good.

Well, I got young kids at home and I tell them, you go to school, you learn. Because the mill is not there to take up for you anymore. That was the people that didn't go to school, they could go to the mill and make as much money as professional people. I quit work, I was makin' 40,000 dollars. The professionals weren't makin' 40,000 dollars in 1980. Nothing more to say, except like I said, I enjoyed workin', I hated to quit workin'.

Pennsylvania and Lake Erie Railroad (P&LE)

Aliquippa and Southern Railroad (A&S)

Engine in scrap alley

A tiny compact fellow with huge forearms from years of track work
and electrical engine repair who wove a tale of a successful life as a self-
educated craftsman.

My name is Steve Palambo. I'm sixty-nine years old and was born
in West Aliquippa. I worked for thirty-eight years and two months in
the Aliquippa and Southern Railroad and was hired in 1946, after the
war. At that time, I was lucky if I made 2,000 or 1,500 dollars a year.
I retired in 1984. Averaging without no overtime, basic pay a year
would be around 17,000 dollars. I worked for the A&S, which is the
Aliquippa and Southern Railroad.

A&S Railroad was a subsidiary of J&L, but they would build up
freight whatever they were shipping out, coils or what. The Aliquippa
Railroad would take it out on the outbound, that's what they called,
which would've been out towards P&LE, and would pick it up for freight
and haul it wherever it goes. So A&S Railroad would take finished
product and then connect to the P&LE, which was the Pennsylvania
Lake Erie Railroad, and then they would ship it nationally.

I started out as what you would call the Maintenance of Way Department, where we repaired track. Well, back when I started they had no heavy equipment. They only had one crane and that was the amount of our heavy equipment. If we got a load of creosote ties shipped in, we unloaded them by hand. We loaded rail onto what they called buggies run on a rail by hand. Everything was done by hand. A railroad tie would weigh a couple hundred pounds. The rail would be, like, if we use ninety pound rail, you are talking ninety pounds to a foot, and there are like thirty-three foot rails and some are thirty-nine foot long.

You would have a crew, let's say, of approximately fourteen men with two men tongs. You picked the rail up and after that we used to drag them. We wised up a little bit and we got pieces of pipe and put them under the rail and roll them on the pipe until later on in years. Many years later, we started getting rubber tire hi-lifts and everything was handled by rubber tire hi-lifts.

So in the early days then you lifted these things yourself so you had a working crew of fourteen guys — yeah, fourteen or fifteen men in a gang. You had about roughly seven or eight gangs. They would've maintained the right-of-way, any wrecks we would repair the track. Any track that was worn, we replaced it with a new track. Plus, in the winter when it snowed and the switches got bogged down with snow, we had to go out and clean and sweep them so that they were workable. From 1946, we cleaned switches all the way to the year I retired in 1984.

From Maintenance-of-Way, my department, I became an operator, what they called track equipment operator, which ran at that time we called them gasoline buggies — motor car. They ran on a rail and they had a Chevy engine in them, like a little car that ran on the track. I did that for roughly the last twelve years, which I became a mechanic and finished up as a lead mechanic. Repaired the stuff I used to run. Repaired hi-lifts or trucks, whatever we had. Towards the end of my career there, we had, like, seven hi-lifts and about ten trucks. We were modernized then.

The locomotives and engines when, I first started, these were steam powered. Steam power engines, you had a engineer and a fireman. The fireman fired the big boiler up with coal to build up the steam and they run by steam. I forget when we had our first diesel.

They were all good to me. It gave me a good wage, good living, and wound up with a decent pension. So I have no complaints about it as far as working down there.

Well, bad days were when it was cold. You had no place to go into so you built what we called a coke-jack, which most of the people in the mill did. This is a fifty-five gallon can, you punch holes in it, use coke to heat it up — that was our bad days. Always outside, until I became a mechanic, and then I worked in the garage, which wasn't what you call a modern garage. Cold, some days it was below zero, and it's hard to work then.

I enjoyed working for the railroad and in the mill. You were meeting friends. We had mostly colored people and white people worked the track gang, which were tie ins and cutters. We got along good. They were good workers, the whole gang. You had to work for a living then. As far as the work gangs were concerned, these were ethnic people basically. Your brakemen and engineers mostly came from the south, were American born.

No, no I miss the people I work with. I remember quite a few of them, quite a few of them died now. I still have about four or five that are still very close, I see. That, and some moved out of town. The younger ones got hired right before the mill shut down. I sort of forgot about them, didn't know them that well. This one got a job, someone else went out of state.

Well, those days it was more fun than work. You went down there you had a good time, nobody got angry at each other. Like, today it's all like what can I get, what do I have to do, it's just different altogether. We look forward to going to work. It wasn't a job that we didn't want to go.

From what I hear it's changed quite a bit. Well, I mean, when I first got a job on a railroad. I would see a fellow working and I would ask a guy how old he is? He would tell me fifty years old and I considered him old at the time. So we would actually try to help him. When I became fifty and the younger fellows were there, they would say let the old guy work, he doesn't have that much time. It just made a flip flop.

Steady daylight. When I first got a job on there, I worked six days a week, Monday through Saturday, we had Sunday off. As the years went by we started working five days a week and was paid for the six, which was Monday through Friday and had every Saturday and Sunday off.

No, I don't think there was anything could change at the time. When I first got a job we didn't have the equipment and, as it builds up, they bought everything they could buy to make it easier for everybody. At the end it was terrific, very good. I retired before the shut down then.

Right, I took an early retirement. They were signing a new contract and half the union went along with the contract and they would let five people retire, the five oldest ones there in my department, and I was the second oldest so I took early retirement. I was fifty-eight when I left the mill with thirty-eight years and two months service. Yes, and I enjoyed the last eleven years of it very much.

About the only thing that I can tell them is, hey, they paid you eight hours so you give them eight hours work. You're not going to go there and sit around, they're not going to pay you for nothing. That's why they're hiring part-time today and getting rid of everybody. All I remember about the early days is hard work.

When I worked on the railroad, if you were an Italian you worked in a track gang. If you were a cake eater — I hate to use the word — I said you run an engine like you're doing now, and it's still that way. 'Cause if your father was an engineer, and you're American, then you're an engineer running an engine. So there's no big change.

Rolls for railcar shipping

Aliquippa and Southern Railroad office

Last railcar

Section VI

The Final Chapter

Chapter 29

ଚଠ୍ଡ

History of Aliquippa

At the turn of the century, 620 people lived in a community called Woodlawn, PA. In 1958, there were 26,369 people living and working in the thriving town of Aliquippa. Jones and Laughlin employed 15,000 workers and the unemployment rate was five percent in the community. Franklin Avenue was a thriving community commercial district.

In 1996, there are 13,374 people living in Aliquippa and over half have moved away since the mill shut down. Currently, LTV employs 700 people and the unemployment rate is twenty percent. Franklin Avenue, the main artery of this once thriving community, is a ghost town.

Only with a great deal of dedication, enthusiasm, effort can this community restore the pride and dignity of "This American Steel Town"

There is as much a need for a strong union in 1997, as there was in 1937 when the National Labor Relations Board vs. J&L decision was rendered maintaining collective bargaining. The methods have changed from strong arm tactics to corporate downsizing, at-will contracts and maximizing shareholder value. Although the methods of intimidation have changed, the effect on the working class has remained the same, decrease in self-worth and respect.

The Mill

The majority of social ills of this country can be attributed to the vast difference between the haves and the have-nots as evidenced by financially decimated communities. The balance between labor and management is not just desirable, but is essential to a productive economy with neither group obtaining the upperhand. A single worker, whether they be a laborer, white collar employee, doctor, or lawyer, does not have a voice as a single individual. The only power the American worker has rests in their numbers, collectively bargaining as a union to maintain "a fair day's work for a fair day's pay."

However, the union is only effective if it maintains its right to strike over issues that adversely and directly impact on its workers, giving up the right to strike finds little help with arbitration and mediation. Only then can it restore the proper balance between labor and management so that the average American worker can maintain a home and family at the dawn of the new millennium.

Rade B. Vukmir, MD
1996

Lower end of Sheffield Avenue

Franklin Avenue with residential housing

Swimming pool with mill relief background

Parade at the "Y"

United Steelworkers of America

The parade route

Upper Franklin Avenue

The grandstand

The parade

Chapter 30

ഇരുള

Epilogue

Aliquippa Business District

Franklin Avenue (1996)

Aliquippa in the fall of 1996

Row houses along old parade route

Parade perch along old parade route

Henry Mancini Storefront

The old Pittsburgh Mercantile

Eger's Jewelers

Florenza's barber shop

Franklin Avenue to the "Y"

The Blue Bell diner

Post shutdown tunnel

Trust in the Lord

References

1. J&L Team Almanac. Volume 29/No. 3, May-June 1976, pgs 9-18.
2. AISI Committee on Manufacturing Problems. Jones and Laughlin Steel Corp Aliquippa Works. 1960, Pgs 3-4.
3. The Making, Shaping and Treating of Steel. JM Camp and CB Francis, 6th ed. United States Steel Co. Pittsburgh, PA. 1951.

Index

Page references followed by *f* indicate figures.
References in *italics* indicate photographs and other illustrations.

Abbott, George, 295
accidents, 11–12, 398; Blast
 Furnace, 93, 103–4; Boiler
 Shop/pipefitters, 247; Carpenter
 Shop, 265–67; crane, 4, *6*, 12,
 195, 343; electrical, 275; falls
 into river, 398; galvanizer fire,
 160; Garage, 288; Machine
 Shop, 316–17; 14" Mill, 142;
 Rod and Wire, 232–33;
 Seamless Tube, 198–200;
 Steelworks, 60–61, 68–69;
 supersucker, 398; Tin Mill,
 213, 218; Welded Tube, 156,
 178; with women, 317. *see also*
 deaths
acid-testers, 228
advancement opportunities, 38
African-Americans, *173*, 355. *see*
 also blacks
Aliquippa, PA, *27*, 93, 122, *216*,
 422; business district, *421*;
 Columbus Day parade, *135*;
 Franklin Avenue, 93, *135*, 143,
 171–72, *216*, 391, 415, *417*,
 419, *422*, *426*; history, 415–20;
 parade route, *419*, *422*; parades,
 63, *193*, *418–420*, *423*; Plan 1,
 341; Plan 2, 341; Plan 3, 341–
 42; Plan 6, 49, 171–72, 310,
 341–42, *344*, 363; Plan 7, 342;
 Plan 8, 310; Plan 11, 310, 341–
 42, 363; Plan 11 extension, 342;
 Plan 11 pool, 169, *174*; Plan 12,
 310, 342, 363; Plan 12 Pool,
 171–72
Aliquippa and Southern (A&S)
 Railroad, 402–13, *408–409*,
 412–413; A&S Railroad Gate,
 388; blacks in, 406; office, *413*;
 tracks, 404

Aliquippa Civic League, 170
Aliquippa Country Club, 323–24
Aliquippa Gazette Paper, 171
Aliquippa Hospital, 61, 287, 317,
 396–397, *400*
Aliquippa Indiana Club, 308
Aliquippa Works, 8, 19, 22, 31*f*;
 construction site, *27*; current,
 290; electric power generating
 facilities, 284; expansion, 20;
 foundation excavation, *299*;
 operations, 21; plant entrance,
 29; shutdown, 20, 179–80, 258–
 259, 289, 359. *see also specific*
 departments, mills
Allegheny General Hospital, 317
Amalgamated Association of Iron,
 Steel and Tin Workers, 307–8
Ambrose, Frank, 199
ambulance calls, 286–87
American Federation of Labor, 18
American Iron and Steel
 Association, 18
American LeFrance Firetruck, *292*
Andeko, Johnny, 248
Anders, Al, 168
angle iron, *150*
Anglos, 353
Annex Building, 331
annual picnics, 266
apprentices, 201, 245, 364
apprenticeships, 355; bricklayers,
 256; Byproducts, 38; Carpenter
 Shop, 262–63; Machine Shop,
 301; 14" Mill, 149; Steelworks,
 67–68, 70; Tin Mill, 217
A&S. *see* Aliquippa and Southern
asbestosis, 257
A&S Railroad Gate, 388
Atkinson, John, 26

Awad, Albert, 255–58
awards for suggestions, 120

B. Lauth & Brother, 30
Babich, John, 307
back injuries, 177, 199–200
Baljak, Ann, 348–51, *349*
ball-busters, 115
Bar and Billet Mill, 19, 120, *124*;
    decline, 22; equipment data,
    125–26; operating statistics,
    125; overhead crane collapse, 4,
    6; production data, 125; repair,
    151
barge docks, *291*
14" Bar Mill, 13, 19, 140–54, *145*,
    154, 167, 211, 289, 343; blacks
    in, 406–7; construction site,
    144; decline, 20; equipment
    data, 153; main parking lot,
    146; production data, 153;
    resumption of production, 20;
    tunnel, *146*; workers, 355
Barr, Paul, 179
baseball, 307–8
Basic Oxygen Furnace (BOF), 20,
    56, 65–67, 72, 74–75, *86*, *89*,
    265; accidents, 275; decline, 22;
    equipment data, 88; operating
    statistics, 88; production data,
    88; shutdown, 20, 289; thefts,
    388–89
basketball, 353
Beaver County Development, 367
beehive ovens, *52*
Begg, Mike, 130
Bell, John, 172–73
Benzol Lab, 43
Bessemer: equipment data, 88;
    operating statistics, 88;
    production data, 88
Beukovic, Anka, 3, *7*
Beukovic, Rade ("Jedo"), 3–4, *7*
"B.F. Jones" (steamer), 19

Biega, Louis, 26
Billet Mills. *see* Bar and Billet Mill
Bisotti, John, 295
Biss, Joe, *119*, 119–22
blacks, 77, 166–68, 170, 172, *173*,
    341, 363, 406–7, 411; in
    railroad, 406; treatment of, 48–
    49, 51
Blacksmith Shop, *325*
Blast Furnace, 4–5, 8, 12, 19, 72,
    90–112, *94*, *96*, *105–106*, 217,
    252, 265–66, *334–335*;
    Christmas star, 15; dangers,
    103; decline, 22; demolition,
    22, *106–108*; equipment data,
    109–10; maintenance, 331;
    preheat, 267; production data,
    109; rebuilding, 256, *260*;
    repair, *326*; scaffolds, 269,
    269–270; shutdown, 20, 93,
    104, 289, 366–67; workers,
    355. *see also specific
    departments*
Blinkey, George, 267
Blooming Mill, 4, 8, 12, 58, 113–
    26, *121–122*, 167, 192;
    accidents, 11; demolition, *124*;
    Engine Room, 120; equipment
    data, 125–26; labor force, *116*;
    operating statistics, 125;
    production data, 125;
    rehabilitation, *117*; shutdown,
    116; slab yard, 11; wash house,
    11
44" Blooming Mill, 10; decline, 22
45" Blooming Mill, 10;
    construction, *121*; decline, 22
blooms, 189
Blue Bell diner, *426*
Blue Ribbon Steel Pipe, 133
BOF. *see* Basic Oxygen Furnace
B.O.F. & Continuous Caster, 20;
    decline, 22; shutdown, 20

Boiler Houses, 243–54, *251*, 321;
flood, 249
boilermakers, 68, 243, 266
Boiler Shop/pipefitters, 243–54. *see
also* pipefitters
bonuses, 68
Boot Straps (Mark), 386–87
bowling league, 233
Boyd, Roselle, 78
braising support, *328*
brakemen, 406, 411
Breedlove, Seth, 147
Breslin, Tom, 26
Brewery Gate, 388, 392
Brick Department, 255
bricklayer helpers, 256
bricklayers, 255–60, *259–260*
Brown, Byrd, 171
Brummitt, Mike, 305
Brummitt, Morris E. (Moe), 25, 232
Brunner, Carl, 307
buckers, 245
Buday, Bob, 157
buddy relief, 157
Bull Gang, 199, 263, 322, 339–40,
343
Burns, Joe, 199, 274
Buttweld Furnace, 364
buyout, 290, 368–69
Byproduct Gate, 388, 392
Byproducts, 8, 34–53, *39–40*, *46*,
211, 249–50, 256, *291*, 325;
riggers, 333; scaffolds, 263–64,
333; shutdown, 20; workers,
355
Byrd, Carrie B., 50
Byrd, James (Jim), *47*, 47–51

Cable, Jack, 267
caddies, 323
cafeteria, 15
cake eaters, 412
camaraderie, 393

carbon monoxide exposure, 15
carpenters, 233, 264, 266, 355;
black, 406
Carpenter Shop, 261–70, 353;
annual picnic, 266
Carr, John, 217–19
Cast House, 14, *95*, *99*
casting process, 100–101
catchers, 245
Cavoulas, Nick, 8, 116
Central Pump House, 254
Charging cranes, 67
Chem Lab, 287
Children's Hospital fund, 357
Christmas star, 15
cinder snappers, 101
Cindrich, John, 374–77, *375*
CJ Betters, 367
cleaners, 321
Clean Sweep, 306
clean-up detail, *299*
clerical work, 176, 227, 348–52,
352
Cleveland Works, 62, 290, 313
clothing, 9; footwear, 9; formality,
363; gloves, 250; heat-
protective, 14; J&L jackets, 84;
protective, 248; winter
underwear, 342; wooden shoes,
264; yellow clothes, 37
Coal and Iron police, 168, 386–87
coal miners, 60
coat hangers, 279
Coe, James R. "Hillbilly Jim," 140–
44, *141*
coiler operators, 279
coil handlers, 340
coke, *52*
Coke Battery: decline, 22
coke "beehive" ovens, *52*
Coke Department: shutdown, 366
coke guides, 48
coke-jack, 411

Coke Plant, 47–48, 53; blacks in, 406–7; description of facilities, 53; equipment data, 53; operating statistics, 53; production data, 53
cold rolling, 18
Cole, Bill, 48
Colona, Guido, 294, *295*
colored people, 411. *see also* blacks
Columbus Day parade, *135*
communications, 275
company credit, 9
company store. *see* Pittsburgh Mercantile Store
Congress Republic Steel, 70
Constitution Boulevard, 15
Continuous Weld (CW) Mills, 175, *187*, 364; equipment data, 185–86; production data, 185
craftsmen, 243
crane followers, 10, 176
crane hookers, 160
crane inspectors, 342
cranemen, 65, 70, 131, 281–82, 341; deaths, 12, 334–335, 343, 377
crane operators, 10, 67
cranes, 340, *345–346*; accidents, 4, 6, 12, 195, 343; Electric Crane Department, 339–46; Electric Crane Repair, 341
Croatians, 60, 363
Crow Island, 389
Crucible Steel, 82
Cubans, 255
Cumberledge, Elmer, 155–57, *156*
cutters, 411
Czechoslovakians, 243, 248, 363

danger, 333–34
Darroch, Patrick A., *65*, 65–73
Davies, Chuck, 305
Davis, Bob, 172–73
Davis, Ed, 351

deafness, 13–14
deaths, 376, 381, 397–98; carpenters, 265–67; cranemen, 12, 334–335, 343, 377; Garage, 287–88; pipefitters, 247; Strip Mill, 130. *see also* accidents
departments, 353–55
Depression, 56, 58, 122, 161, 279
depression, 326
DeSena, Blackie, 313, 358
DeSena, Louis, 26
DeVaney, Paul, 323–24, 388
disability compensation, 200
discrimination, 37, 48–49, 58, 77, 115, 166–68, 171–72, 308; against blacks, 406–7
dogs, 246
Dolnack, Mikey, 70
Downing, James (Jim), Jr., *167*, 166–73
Downing, James (Jim), Sr., 166–70
drinking, 39
drivers, 212, 245, 286, 294–95
Duckworth, T.D., 8, 17
Dudak, Alex, *192*, 194–97
Dugas, John, 248
dump trucks, 286–87
Duquesne Light Company, 244–45
Dye, Roy, 49–50, 170
Dzumba, Johnny, 219

ear plugs, 250
Eger's Jewlers, *425*
Electrical Department, 198, 271, 273
electrical power generating facilities, 284
electrical power supply, *283*
Electric Crane Department, 339–46
Electric Crane Repair, 341
electricians, 37–38, 149, 201, 217, 233, 271–84; blacks, 406; Master Electrician, 272–73; motor inspector, 280; shop, *280*

Electric Weld Mill, 163–65, 175,
188; equipment data, 186;
production data, 186
Electrolytics Department, 285
Electronics, 272
Eliza furnaces, 18
Elks, 301–2
emergency care, 287–88
employee stock ownership (ESP),
358
engineers, 120, 410–12
Eritano, Peter J., 26
ESP. *see* employee stock ownership
ethnic groups, 326–27, 411;
groupings, 353–55

family life, 177
fatalities, 376, 381, 396–98;
carpenters, 265–67; cranemen,
12, 334–335, 343, 377; Garage,
287–88; pipefitters, 247; Strip
Mill, 130. *see also* accidents
Ferezan, George, 209–15, *210*
fiberoptic lung disease, 13
fighting, 172
Finishing Mill operators, 279
Finishing Mills, 11, 120
firemen, 406, 410
Firestone, 142
first helpers, 102
Flat Mill, 192
Flexco plates, 43
floods, *26*, 249
Florenza's barber shop, *425*
football, 308
footwear, 9; wooden shoes, 264
Forbes, Jim, 130
Ford, Gerald, *373*
foremen, 334–335, 374–75, 405
Franklin Avenue (Aliquippa, PA),
26, 93, 143, 171–72, *216*, 391,
415, *417*, *419*, *422*, *426*;
Columbus Day parade, *135*

Frioni, Dennis, *147*, 147–50
Frost, Bill, 14
Fry, George, 301
Furher, Vince, 116

Galvanized Department, 164, 179
Garage, 285–93
gas masks, 248
Gear, John, 50
General Electric (GE), 163, 176,
356–57
general grievers, 353, 355, 365
General Labor, 74, 294
General Maintenance, 242–346;
shutdown, 289. *see also specific
departments*
Geneva, 278
George, Alex, 26
Germans, 66
Gestetnners, 348
Giammaria, Joe, 276
GI Bill of Rights, 304–5
girder erection, *338*
gloves, 250
golf, 233
Gomer, 11
Goonis, Al, 60
Graham, Tom, 62
Greeks, 353–55
grievances, 271–72, 356–57
Grossi, Jimmy, 60
guards, 388, 391–92

Hallisey, Pat, 388
hand injuries, 213, 218
harassment, 306, 308
Hatfield, Dan, 232
Hayes, Richard, 18
Headquarters Gate, 388
healthcare, 61, 130, 313, 397;
physical examinations, 257
Health Center, 130, 287, 395–401

health hazards, *401*; blisters, 57–58;
Boiler Shop, 249–50;
bricklayers, 257; Byproducts,
37, 40, 44–45; 14" Mill, 150;
Steelworks, 60; Strip Mill, 130–
31. *see also* accidents
hearing aids, 250
hearing loss, 13–14, 160, 165, 201
heaters, 245
Heating Room (Byproducts), 48
heat-protective clothing, 14
hectographs, 348
Henderson, David, *74*, 74–84
Hennepin Works, 20, 62, 313
Henry Mancini store, *424*
high flyers, 267
High Mill, 195
holidays, 164
Holmes, Ben, 179
hopper cars, *96*
Hot Bed, 120
hot bed operators, 141
Hot Mill, 59, 179, 194–95;
furnaces, *206*; Old Hot Mill,
212; Reducing Hot Mill, 129
44" Hot Strip Mill, 273
Hot Strip Rolling Mill, *134–135*
Howell, Jim, 147, 149

Iannessa, Lelio, 280–82
I-beams, 13, *150*; Junior I-Beams,
145, 148
IBM, 348
ICS. *see* International
Correspondence School
immigrants, 326–27
industrial nursing, 398–99
injuries, 232, 286; crush injuries,
397; hand injuries, 213, 218;
traumatic injuries, 398. *see also*
accidents
Inman, Donald, 271–79, *272*
Inspection Department (Seamless
Tube), 194

Inspection Department (Welded
Tube), 163, 175
inspectors, 175–76
insurance, 57
International Convention, 257
International Correspondence School
(ICS), 262–63
Italians, 37, 341, 354, 363, 412

"James Laughlin" (steamer), 19
Japan: unions, 359
job classification system, 54–55
Johns, Dave, 306
Johnson, Lyndon Baines, *372*
jokes, 179, 200
Jones, Benjamin Franklin, 18, 30
Jones, Benjamin Franklin, Sr., 19
Jones, BF III, *381*
Jones, Lauth & Company, 18
Jones & Laughlin, American Iron
Works, 18, 30
Jones & Laughlin's, Limited, 18
Jones & Laughlin Steel Company,
17–32, 122, 130, 147–48, 150–
151, 210, 230, 289, 309, 342,
344; annual production, 18;
chronological history, 18–20;
expansion, 19; history, 30;
incorporation, 19; jackets, 84;
merger with LTV, 290, 368–69;
National Labor Relations Board
vs. J&L, 415; ore properties,
18; product line, 20; site, 367.
see also Aliquippa Works
Jones & Laughlin Structural Steel,
20. *see also* Jones & Laughlin
Steel Company
Junior I-Beams: production, *145*,
148
Jurasko, Bob, *175*, 175–82

keepers, 102–3
Kennedy, John, *199*, 198–201
Kennedy, John F., Jr., 178

Ketchum, Mike, 48
Kier, S.M., 18
Koolaid, 179
Kosanovich, Dan "Spider," 339–44, *340*
Kosanovich, Melvin, 353–60, *354*, 391
Krivan, Joseph, 26

Labor Day celebrations, 226, 231, 233, *237*, *314*
labor gangs and laborers, 14, 294–99, *297*, 353; blacks, 406; Blast Furnace, *98*, 217; Blooming Mill, 113; clean-up detail, *299*; Electric Crane, 340; General Labor Department, 74, 294; Mechanical, 323–24; Rod and Wire, 227, 230; skilled labor, 302; Steelworks, 54, 56–58, 60, 75; Strip Mill, 127, 132; Welded Tube, 156, 166, 176; women, 320
labor union. *see* union
Lackovich, Nick, 360
ladles, 100; submarine iron, *99*
Lane, George, 306–7
Lang, Grace, 398
Lasting Mill, 164
Laughlin, Charlie, 466
Laughlin, James A., 18, 30
Laughlin & Company, 18
Laughner, Charles, 171
Lauth, Bernard, 18
Lauth, John, 18
layoffs, 282, 326–27, 393
lead mechanics, 410
Lebanese, 363
Leechman, Steve, 334
Letteri, Joe, 267
Letteri, Mario, 267
"Liars, Thieves, and Vultures," 73
Ling, John, 368–69

Ling-Temco Vought, Inc. (LTV), 20, 38, 46, 62, 130, 150, 210–11, 278, 280, 313, 342, 344, 367, 376, 415; as "Liars, Thieves, and Vultures," 73; merger with J&L, 290, 368–69; purchase of Youngstown Sheet and Tube and Republic Steel, 133
Logstown, 341, 363, 388, 391–92
LTV. *see* Ling-Temco Vought, Inc.
lunch breaks, 131

machine maintenance, 130
Machine Shop, 5, 300–322; North Mill, 300; South Mill, 300–301
machinists, 301–2, 315
maintenance, 130, 242–346
Maintenance-of-Way Department, 410
Mamula, Nicholas (Nick), 25, 226, *371*, 375
management, 181, 374–85, *383–384*; safety rally, *383*
management office, *385*
managers, 50
Mancini, Fred, 34–40, *35*
Manolovich, Nick, 232–33
Maravich, Bronko, 301, 305–6
Mark, H.G.: *Boot Straps*, 386–87
married women, 321
Master Electrician, 272–73
Matish, Eli, 113–16, *114*
Meaney, George, *372*
Mechanical Department, 323–29
mechanics, 286, 410
medical car, 286
medical center, 61
metallurgical engineering trainees, 378
metallurgical investigators, 378
midnight shifts, 157
14" Mill Motor Room, 284

millwright helpers, 43, 324
millwrights, 43, 45, 233, 323–24, 329, *327*, 342
Milne, Bill, 393
Mitchell, Todd, 169
Mitko, Larry, 285–91
mixermen, 78, 90
molybdenum: theft of, 388
Monarch Steel Company, 20
Morell, Ben, *380*
motor inspectors, 36–37, 198, 218
Mrogan, Tom, 195
Musolin, Sam, 391

Nail Mill, 167, 233, 240
National Electric, 163
National Labor Relations Board vs. J&L, 415
National Recovery Act, 57, 121
National Recovery Administration (NRA), 57
Nellish, John, 307
New Dravo-Lurgi Sintering Plant, 13, 90, 91–93, 110–11, *112*, 332–33, 340
night shifts, 157, 317
Normile, Paul, 26
North Mill, 13, *28*, 33–138; Boiler Houses, 253; Machine Shop, 300; Pump House, 254; riggers, 332–33. *see also specific departments*
North Mill Relief, 392
North Powerhouse, 284
NRA. *see* National Recovery Administration
nurses, 287, 395, *396–397*, 398–99

Occupational Health and Safety Administration (OSHA), 249
octopus, 246
oil crisis, 312–13
oilers, 120, 301

oil greasers, 43
Old Hot Mill, 212
Open Hearth Furnace, 55–56, 59, 62, *64*, 100, 167, 256, 390; breakdown, 230; equipment data, 88; operating statistics, 88; production data, 88; sample, *63*
Operations Management (Tin Mill), 378–79
order clerks, 176
ore yard, *105*, *345*
Otis Steel, 62
outside contractors, 286–87
overtime, 164, 177, 190, 402

Paige, Emil, 343
Paik, Yung, 149
painters, 343
Paint Shop: workers, 355
Palambo, Steve, 409–12
Palumbo, Augie, 8
parades, *418–420*, *423*; Columbus Day, *135*; golden jubilee, *63*; jubilee, *193*; route, *419*, *421*
Parducci, Johnny, 334
Pastine, Bill, 304
Pastine, Phil, 308
patriotism, 11
pattern makers, 264
Patterson, Red, 267, 333
Paul, Janice, 395–400, *396–397*
paydays, 368
pay wagon, 277
Pennsylvania and Lake Erie (P&LE) Railroad, 403, *408*, 409
pension fund, 70
pensions, 132, 196, 290, 313, 342, 366, 405
Perriello, Joseph (Joe), 300–313, *301*, *314*
physical examination, 257
picnics, 266

Pike, Charlie, 45
pipefitters, 44, 199, 213, 243–54, 266, 364
pipe inspectors, 177
Pipe Shop, 322
Pipe Shop Champion superteam, *251*
pipe threaders, 160
Pittsburgh Mercantile Store, 84, 93, 170, 310, 368, *424*
P.M. Moore Company, 287, 368
P&M Store. *see* Pittsburgh Mercantile Store
Poles, 127
police, 392; weapons, 390
Police Department, 386
police sergeants, 387, 393
Pollick, Mike, 26
Prajsner, Joseph, 132
Prajsner, Marion, 127–33, *128*
preheat operators, 46
Production Planning Department, 348
protective gear, 248, 250; gloves, 250; heat-protective clothing, 14; respirators, 13, 37; safety belts, 267; safety glasses, *318*; wooden shoes, 264
Pudyh, Kazimierz (Kaz), 90–94, *91*
pump man, 120

Quarterback Club, 229

Rabbi, 276
racism, 308, 406–7. *see also* discrimination
railroad. *see* Aliquippa and Southern Railroad
rebar work, 264
Rebich, Michael, 226, *227*
Red Cross, 230
Reducing Hot Mill, 129
Reed, Ed, *42*, 42–46
repair shop, 356

Republic Steel, 20, 72, 73, 133, 289, 369
Republic Steel Plant, 133
respirators, 13, 37
retirement, 74, 230; early, 74, 114–15, 132, 182, 201, 250, 290, 296, 330, 342, 348, 392–93, 405, 412; packages, 232; pension fund, 70; pensions, 132, 196, 290, 313, 342, 366, 405
Reversing Mill, 129, 275
riggers, 3–4, *5*, 57, 266–67, 330–38, 353
right to bargain, 196–97
right to vote for union, 309
Rinaldi, Tony, 313
Ringer, Ed, 247
river: falls into, 398
riveters, 245, 248
Rivetti, Anthony, *162*, 162–65
Rockavich, John, 213
Rod & Wire Mill, 19, 62, 132, 226–41, *228*, *234–236*, *241*, 279; decline, 22; demolition, *236*; equipment data, 237–40; production data, 237; shutdown, 20; workers, 353
roller scale repair, *329*
roll grinders, 127
Rolling Mill: labor, *297*; repair, *136*
Roll Shop, 127, 130; accidents, 213
Rosa, Joe, 332
Ross, Carl, 261–69, *262*
Rossi, Joe, 393
30" Round Mill, 189, 192, *207*, 325; decline, 22; equipment data, 204; production data, 204
Russians, 341

safety, 101–2, 131, 249–50; crane, 341; management rally, *383*
safety belts, 267
Safety Committee, 288

"safety first" programs, 19
safety glasses, *318*
safety men, 4
salary, 374
Salvoldi, Louis, 189–92, *190*
saw grinders, 261–62
Saxler, Harry, 388
scaffold builders: high flyers, 267
scaffolding, 263–64, *270*, 333
scarfers, *118*
schedulers, 176, 179
scheduling, 36–37, 45
Schultz, Regis, 360
scrap alley, *409*
Seagrave Sedan cab, *292*
6" Seamless Mill: equipment data, 204–5; production data, 204
14" Seamless Mill: equipment data, 204; production data, 204
Seamless Tube Mills, 8, 19, 142, 189–208, *197–198*, *202*, *206*, *208*, 324–25; decline, 22; equipment data, 204–5; floor rehabilitation, *193*; former site, *203*; inventory, *202*; jubilee parade, *193*; laborers, 353; oil crisis and, 312–13; production data, 204
second helpers, 102
Security Department, 386–94
segregation, 48–49
Serbians, 59–60, 159, 226, 229, 341, 343, 353, 363, 390
service, 261, 347–413
severance pay, 326–27, 392
Sheffield Avenue, 416
shift work, 142–44, 160, 177, 180, 195, 249, 288–89, 326; midnight shifts, 157; night shifts, 317; scheduling, 36–37, 45; swing shifts, 121
Shipping Yard, 364
shooting range, *394*
silicosis, 257

Sims, Jim, 78–79
Sinter Plant. *see* New Dravo-Lurgi Sintering Plant
skid boys, 212
skilled labor: standardized, 302
skip car, *97*
Skorich, George, 268
Slab Yard, 120
slag alley, *97*, 101, *298*
Slag Disposal Gate, 388
Slavs, 58
Slovaks, 58
small bell, *336*
Smith, Dave, 93
Smoljanic, Brownie, 332, 334
Snyder, Carl, 147, 149
Soaking Pit, 115, 132, 342
SOAR. *see* Steelworkers Organization of Active Retirees
social life, 60
softball, 230
South Heights Gate, 388
South Mill, 13, 139–241, 377; Boiler Houses, 249, 253; Machine Shop, 300–301; Pump House, 254; riggers, 332–33. *see also specific departments*
South Mill Overhead Bridge Gate, 388
South Powerhouse, 284
South Side Hospital (Pittsburgh, PA), 61, 130
South Side Lauth & Company, American Iron Works, 30
sports: baseball, 307–8; basketball, 353; football, 308; softball, 230
Stasko, John, 248
Stauffer, George, 390
stealing, 172, 357, 388–90
steel: cold rolled, 18; erection, *338*; hot rolled bar, 18; production, 19, 55–56; sorting and stocking flats, *145*; structural, 148

steel cast, *72*
steel pourers, 68
Steelworkers Organization of Active
   Retirees (SOAR), *16*, 182
Steelworkers Organizing
   Committee, 307
Steelworks, 54–89, *71*, *86*;
   construction, *85*; labor gang,
   85; shutdown, 366
steel yards, *152*
Stephans, Bill, 378–81
Stephens, Bill, *98*
Stetz, George, 307
Stevenson, Lewis, 172
Stockhouse, 12, 77, *97–98*; control
   room, *98*
stokers, 406
Stokes, George, 402–7, *403*
stool-pigeons, 303
stove men or stove tenders, 102–3
Strandcaster, 56, 65–66, 192
Strand Theater, *215*
Strip Mill, 127–38, 271, 273;
   demolition, *137*
44" Strip Mill, 129, *137*; decline,
   22; equipment data, 138;
   production data, 138; Roll
   Shop, 129
Strothers, George, 296
Strothers, John, 295
14' Structural Mill, 129
structural steel, 148
subforemen, 38
Sudar, John T., 386–93, *387*, 391,
   393
superintendents, 130, *382, 384*
supersuckers, 398
supervisors, 393
Sweringen, Wayne, 179
swimming pool, *417*
swing shifts, 121

table operators, 190
Tandem Mill, 209, 218
"tapping the hole," *95*
Teleha, Michael, 54–62, *55*, 68
Temple Theater, 214
temporary employees, 325
tester operators, 325
thefts, 172, 357, 388–90; of
   molybdenum, 388
Thomas, Floyd, 219
tie-ins, 411
Tilter Building, 92
tin floppers, 321
Tin Mill, 8, 59, 70, 113, 116, 209–
   25, *217, 220–221, 225*, 280,
   289, 312, 325, 340–41, 343,
   364; blacks in, 406; clerical,
   348; current operation, 351;
   employees, 355; equipment
   data, 222–24; inventory, *220*;
   Operations Management, 378–
   79; production data, 221;
   products division, 379;
   resumption of operation, 20. *see
   also specific departments*
Tin Plate Department, 19; decline,
   20
track equipment operators, 410
track gangs, 411–12
track men, 403–6, 410
tractor drivers, 212
train crews, 406, 410
training program, 378
Tremont, Ray, 393
Trotta, Chris, *315*, 314–18
truck drivers, 286, 294–95
trucks, 285–86; American LeFrance
   Firetruck, *292*
Tube Mill: blacks in, 407
Tube Mill Gate, 388, 392
tunnel, 19, *427*

Turkovich, John, *158–159*, 159–61
turn foremen, 375
typist/clerks, 348

Ukrainians, 113
unemployment compensation, 60, 392
unemployment rate, 415
union(s): current, 289–90; in Japan, 359; old, 290; right to vote for, 309. *see also* United Steelworkers of America
Union Hall, 169, *361*
union man, 328
Unis Trucking, 287
United Sons of Vulcan, 18
United Steelworkers of America, 133, 181–82, 196, 233, *418*; enrollments, 169; membership oath, 24–25; Steelworkers Organization of Active Retirees (SOAR), *16*, 182
United Steelworkers of America Local 1211 (USW-1211, Aliquippa Chapter), 166–68, 249, 258–259, 304, 312, 350, 353–73, *362*; benefits of, 39, 368; committeemen, 61, 69, 255–57, 288; discrimination and, 48; Dues Protest Committee, 375; flag salute, *362*; formation, 358; grievance process, 46, 355–56; grievances, 271–72, 356–57, 391; infant stage, 167; meetings, *363*; membership, 365; negotiations, 365; officers, *371–372*, 391; participation in, 307; power of, 357; presidents, 25, 357, 365; representatives, 365; stewards, 42–46; strike food bank, *370*; strikes, 92, 170–72, 179, 196–97, 305, *337*, 358–59; voting, *363*, 365; wages and, 49; worth of, 313

Urick, Sam, 230
USW. *see* United Steelworkers of America

vacation time, 49, 302, 376
Vallecorsa, Richard, 25, 313, 363–69, *364*
Verez, Curley, 333
Vignovich, Fred, 332–33, *337*, 360
Vignovich, Mitch, 330–34, *331*
Vladovich, Anthony (Tony), 25, 226, 301, 306, 391
"Vote Republican Reagan," 71
voting, 308; for union, 309
Vukmaravich, Lubar, 332
Vukmir, Leni B., 4–5, 15, *319*, 320–22
Vukmir, Rade B., 2, 12–14, *16*
Vukmirovic, Matthew (Matt), 3–4, *5*, 332, 343, 360; death, 12, 334–335, 343, 377

wages, 10; Blast Furnace, 100, 104; Blooming Mill, 121; Boiler Shop/pipefitters, 243; bonus and incentive, 11, 121, 141, 249, 285, 407; Carpenter Shop, 261; cost of living raise, 302; current, 323; grievances, 271–72; hourly rates, 23, 280, 285, 300–301, 314, 323, 340, 348, 364, 374, 386, 394, 402; machinist, 301–2; 14" Mill, 140, 147; monthly, 386; overtime, 402; rates, 37–38, 245; Rod and Wire, 232, 234; salary, 226, 232–33, 271, 374, 395, 409; Seamless Tube, 189; sliding scale, 18; Steelworks, 49, 54, 58, 65; Tin Mill, 217; Welded Tube, 155, 159, 161
Wagner Act, 196–97
Walker, Jughead, 265
war effort, *28*
Warren, OH, 325

water pumping, 254
water tenders, *7*, 103
Welded Tube Gate, 388
Welded Tube Mills, 19, 62, 155–88,
    158, *166*, *174*, *183*, 364;
    decline, 22; equipment data,
    185–86; former sites, *184*;
    production data, 185
welders, 374
West Aliquippa, PA, 363
West Aliquippa Island, *87*
wire, 279
Wire Mill, 312; shutdown, 313
women workers, 77–78, 172, *322*,
    340–41; accidents, 317; clerical,
    348–51; laborers, 320; married
    women, 321
wooden shoes, 264
Woodlawn, PA, 415. *see also*
    Aliquippa, PA
Woodlawn Hotel, 168
workers: employment rate, 415;
    health, *401*; hourly, 376, 392;
    salary, 376; treatment of, 129;
    unemployment rate, 415
working conditions, 46, 104, 131,

150-151, 195, 218, 250, 349;
    Byproducts, 36–37
workman's compensation, 70, 72,
    115, 258
work schedules, 36–37, 45, 93, 156,
    375, 411; holidays, 164;
    overtime, 164, 177, 190, 402;
    vacation time, 49, 302, 376
World Processing, 211

Xerox, 348

Y, 168, *418, 426*
Yakich, Francis, 100–104
yellow clothes, 37
Young, Mike, 250
Youngstown Sheet & Tube, 20, 72,
    133, 289
Yurich, Nick, 343

Zelenak, Martin, 243–50, *244*
Zelenak, Martin, Sr., 248
Zernich, Michael, 200
Zinz, Bill, 307
Zitzman, Francis, 267
Zuccaro, Ronald D., 323–27, *324*

John May
Ed Dulay
Rich Rippee
J.R. Dempsey
Paul Barr

Henry J. Roscoe
Ron. [signature]
Al Jarvis